Recoding
Scientific Publishing

Raising the Bar in an Era of Transformation

Bhakti Kshatriya, PharmD

Dedication

This book is dedicated to my parents, Minaxi & Kishor Parmar, and my best friend, Kanha, for the inspiration.

Special thanks to my husband and our two wonderful daughters for their love and support to make this book a reality!

Acknowledgments

I am sincerely grateful to my husband, Raj, for his moral support, suggestions, book title, and cover design concept. I would also like to thank my friend, Mrunal Deshpande, for helping with book cover design, and to Editor Virginia of FirstEditing.com for her helpful editorial suggestions.

This book was possible thanks to the experiences shared with many friends and colleagues within publication profession and scientific research community.

Contents

Preface

I was introduced to medical writing and publications in 1996 during last year of my doctor of pharmacy program, while attending a job fair at the American Society of Health-Systems Pharmacists (ASHP) midyear meeting. At the time, I knew little to nothing about publication planning or medical writing in order to consider it as a career option or profession. Interestingly, after graduation, I landed my first job as a medical writer at a medical communication agency, which gave me a tremendous opportunity to learn the basics of who, what, when, and how of scientific publishing and medical education. It was my first exposure to publications 101—about congress abstract submission, poster presentation, oral presentations, manuscript development, and submission to journals, as well as the process of peer review, rejection, conditional acceptance, and finally acceptance for publication! This was also my first indoctrination into how industry-sponsored publications are developed, and how the rules and practice vary from company to company. This further paved my professional path and journey that offered me the honor and privilege of working with and getting to know hundreds of fellow medical and clinical professionals within the industry, as well as international medical experts from around the world, for a wide range of therapeutic areas.

Medical writing was my initial learning ground and became my foundation to reflect upon as I grew into the role of publication manager. My first opportunity as publication manager was at Aventis Pharmaceuticals (which was later acquired by Sanofi), specifically for managing development of scientific publications and enduring medical education materials to support their US organization. This was my first experience working directly within a pharmaceutical company, leading publication teams to plan and execute publications and enduring medical education materials generated from US-led company-sponsored studies of established products as well as investigational drugs. I further learned the do's and don'ts of pharmaceutical company engagement with academic researchers as well as US regulatory environment. My next stop was at ALTANA Pharma AG (which was later acquired by Nycomed) as publications leader overseeing publication activities, including internal medical writers, for studies led by their global research and development team. In addition to interacting with associates and investigators worldwide and learning about the European and Asian regulatory environment, this is also where I gained experience in publication planning and delivery to support pre-launch and launch of the company's two leading compounds. Each product had different co-development and co-promotion agreements with two different companies: one was with Aventis Pharmaceuticals specifically for the United States, while ALTANA held the rights for rest of the world outside the United States, and the other was with Pfizer for worldwide co-development and promotion. This allowed me to gain unique experience establishing, co-leading, and managing large publication teams that involved members from two different companies.

After ALTANA, my next adventure was at Novartis Consumer Health (which is now a joint venture with GlaxoSmithKline Consumer Healthcare). This was an unexpected and unique experience to apply the principles of publication and medical communication planning and execution for over-the-counter (OTC) or direct to consumer medicines, including supporting the switch from Rx (prescription) to OTC. All my experience prior to joining Novartis Consumer Health was focused on prescription-based medicines, ranging from pre-launch to launch and on through to mature established products. Thus, working at Novartis Consumer Health allowed me to complete my experience for an entire product life cycle, including the Rx-to-OTC switch.

My dream of contributing to publication and medical communication of hematology and oncology research came true when I joined the global scientific communications group within Novartis Pharmaceuticals Corporation—oncology business unit. This is where I further built on my experience and contributed to both publication and medical education activities across the globe for investigational as well as established products. Leveraging all my prior publication and medical communication experience,

I was able to contribute and lead development of many of the company's publication-related standard operating procedures and guidance, as well as the corporate publication policy. In addition, more than half of my nine-year tenure at Novartis Oncology was dedicated to establishing and strengthening the publication excellence capabilities within the organization with regard to compliance and monitoring of publication practices, setting procedures, training of the internal medical writing group and medical associates worldwide, administrative oversight of publication management system, and health care professional (HCP) payment transparency reporting.

After over ten years at Novartis, I have now founded Publication Practice Counsel to continue to contribute to publication profession by sharing my knowledge and expertise on ethical and good publication practice with pharmaceutical, biotechnology and medical device companies, academic institutions, and researchers in their efforts to publish scientific and clinical research.

Over the last two decades, my passion for publications has continued to grow and I consider myself fortunate and proud of being part of the positive evolution and transformation of scientific publishing. Today, it is a recognized profession within the industry, along with established professional societies and organizations such as the International Society of Medical Publication Professionals (ISMPP), the International Publication Planning Association (TIPPA), American Medical Writers Association (AMWA), and European Medical Writers Association (EMWA). Sometimes, I wonder why I have had continued interest in scientific publishing for such a long time. After deep soul searching, I realized I find profound satisfaction and gratitude in being able to contribute to something that can remain imprinted *forever* to ultimately help *humanity* for generations to come. Unless we have an apocalyptic disaster that brings an end to our world, scientific publications are a permanent record of research, discovery, and inventions. They allow us to take a peek into the historical knowledge possessed by our ancestors at any given point in time, while allowing us to share our current knowledge for future generations.

We are living in an era of monumental transformations. It's an amazing feeling to be blessed to witness some of the most exciting and miraculous changes. One of the most prominent and life changing transformations has been related to how we communicate. As I was growing up, we communicated long distance with our friends and family via a letter or phone call. We can now connect with our loved ones instantaneously via email, texting, phone call, or video call—technology has been nothing short of a miracle and has helped close distances of thousands of miles. I am looking forward to the day when we can all connect with others via hologram and virtual presence!

Scientific publishing has also undergone major cultural and positive transformations over the last decade, particularly in the areas of transparency and access to scientific information. And we should celebrate our achievements. In essence, through incremental changes in our thinking and approach, we are *recoding* the process and culture of scientific publishing and information exchange. Standardization of disclosure of medical writing support, funding source, and conflict of interest in scholarly papers; clinical trial registration and results disclosure on public repositories; and publication of regulatory documents for public access are some examples demonstrating the cultural shift in transparency. Emergence and availability of thousands of open access journals, public repositories that provide free access to published articles, and data sharing initiatives that will provide researchers access to existing data in order to conduct further research and uncover new findings—all exemplify our cultural shift toward free and open access to scientific information. Some of these examples apply to both transparency and access. With any change, there is often pain before experiencing pleasure. And the scientific community is not immune to this either—it endured denial, skepticism, controversy, criticism, and debate over many of these topics before it collectively came to a realization and acceptance of embracing the new culture. I admire, commend, and am deeply thankful to all those who believed in and remained focused and committed to making these positive changes a reality. Their vision and efforts can be summarized in the following words by Mahatma Gandhi:

> *"First they ignore you, then they laugh at you, then they fight you, then you win."*

Being part of this cultural evolution is what inspired me to write this book. I believe that the cultural shift is not done yet; it's only just begun.

Our way of thinking, our mindset, and the way we communicate and exchange scientific research and information will continue to undergo more transformation. Changes in our social culture, norms of communication, social globalization, and newer technology will act as catalysts for further transformation in how scientific publications are compiled and accessed. In my opinion, medical science research and publications are grounded and rely on four fundamental, interdependent, critical elements:

- Scientific Integrity
- Transparency
- Access
- Speed

All of the above critical elements can in many ways be considered moral and ethical obligations to the public and patients, who voluntarily devote themselves by participating in clinical trials for the betterment of the health of future generations. A breach or void in any of the above critical elements can have a direct or indirect impact on *trust*, which is the *life breath* for science and medicine.

Scientific integrity is the foundational element that the scientific research community dealt with early on, at the turn of the twentieth century, when there was no widely accepted code of conduct governing the ethical aspects related to human research. The 1947 Nuremberg Code served as a predecessor to what is currently recognized and followed—the Declaration of Helsinki, adopted in 1964—that provides a set of ethical principles for medical doctors when conducting research in humans. It provides a standard for protecting human rights of those who participate in clinical trials as research subjects. The declaration is one of the most widely recognized, and it forms the basis of additional guidelines such as Good Clinical Practice (GCP). GCP provides international, ethical, and scientific quality standards for designing, conducting, recording, and reporting trials that involve human subjects. All clinical trials are expected to comply with GCP standards.

The International Committee of Medical Journal Editors (ICMJE) first issued its "Uniform Requirements for Manuscripts Submitted to Biomedical

Journals" (URM) in 1978, which initially included guidance on the physical characteristics and format of a manuscript. Over the years, they have evolved to address ethical aspects of publications such as authorship, duplicate publications, conflict of interest, definition of peer-reviewed journal, and requirement for clinical trial registration. The latest version of ICMJE's "Recommendations for the Conduct, Reporting, Editing, and Publication of Scholarly Work in Medical Journals" is available on www.icmje.org.

More recently, Good Publication Practice (GPP) guidelines, first published in 2003, were introduced to elevate scientific integrity of publications and provide ethical principles and quality standards for developing publications that report industry-sponsored research. GPP is now in its third version—GPP3 (September 2015)—and although the guidelines are written from the perspective of industry-sponsored publications, its ethical principles and quality standards can be applied to any research-based publication. Furthermore, the Council of Science Editors (CSE), an international organization of editorial professionals, issued "White Paper on Promoting Integrity in Scientific Journal Publications, 2012 update" that also provides ethical principles and guidelines on roles and responsibilities of journal editors, publication authors, peer reviewers, and study sponsors, as well as on identification and corrective actions for research misconduct. The Committee on Publication Ethics (COPE) is another international organization, established in 1997 with over 10,000 members, primarily journal editors and publishers, that provides advice and ethical standards for editors and publishers. COPE has published the "Code of Conduct for Journal Editors," as well as several guidelines for journal editors-in-chief, peer reviewers, publication editors, and how to handle authorship disputes. Together, quality standards have helped create and establish the culture of scientific integrity in clinical research and scientific publications, which is quintessential for ensuring trust from and within the scientific community and general public.

With regard to transparency and access, the research community and industry underwent public and media scrutiny for lack of transparency and open access to scientific information prior to undergoing the cultural shift. Therefore, the resultant changes can be considered reactive rather than proactive. In my opinion, whenever a change occurs only in response to criticism of a certain practice or behavior, there is the risk of losing the trust, respect, and credibility of those who questioned or criticized the inappropriate behavior. Despite the many positive changes, results of the US General Social Survey did not show an improvement in public trust of science from 1974 to 2010 (Gauchat, *American Sociological Review,* 2012). A multivariate analysis of the data showed a decline in trust of the scientific community among political conservatives. Similarly, according to the Eurobarometer Survey for the European Commission, almost 60% of respondents expressed distrust toward the scientific community (Eurobarometer, "Science & Technology Report", 2010). The pharmaceutical industry continues to face scrutiny and public distrust, probably for the same reasons. One way to avoid losing trust while maintaining respect and credibility is by making the change proactively before giving others an opportunity for criticism. How can we identify if something needs a change? One way is by considering *the right thing to do.* For those of us involved in clinical research, that could be in reference to patients—what is the right thing to do for patients? Amid debate and strong opinions for and against the data sharing requirements proposed by the ICMJE for publications, it's reassuring and inspiring to see the authors of a recent editorial in the *New England Journal of Medicine* refer to the current data sharing movement as "the right thing to do" (Rockhold et al, *NEJM,* 2016). There will always be differences of opinion. However, if we practice and apply the principle of "the right thing to do" in our thinking and decision-making process, it could help keep us from getting into trouble and public scrutiny later on.

The fourth critical element, speed, has not received a lot of attention thus far, but it has a significant opportunity for improvement. Our current mindset and culture have accepted that it is all right to take over two years to get a scientific paper published in peer-reviewed journals. However, two years is a phenomenally long time to get a paper published. In the current era of digitization and technology, shouldn't we be able to publish a lot quicker? Technology companies, including Apple, Google, Facebook and many more, are now tapping into the health care industry; and technology's main mantra is *speed, convenience, and open communication.* The introduction and marriage of technology with medical science is likely to bring some fresh thinking and cultural shifts to elevate the value of time and further enhance openness in communication of scientific research and information.

In this book, we will take a look at the current landscape of publishing life sciences research—who is involved, the process of publication planning, how publications are developed and eventually published—along with presenting helpful tips and suggestions to optimize the publication process within the current infrastructure and landscape. The book also provides insights into the performance and compliance of the research community in dissemination and communication of clinical research within the last decade. Lastly, it gives a vision for the path forward with an overview of anticipated changes in the near future within the scientific publishing environment, as well as thoughts and ideas for longer-term transformation.

The discussion is primarily from the perspective of medical research publishing. The concepts related to the above four critical elements are embedded throughout the book. In order to allow a look at the *current* landscape and situation, an effort has been made to include review of the most recent studies and literature related to various topics on scientific publishing as much as possible. In many areas, there is either lack of, little or emerging reports of formal assessments on certain topics. While it is optimal to draw conclusions based on robust evidence, it is interesting to see some emerging data either supporting or refuting certain perceptions and standard practices related to publications. As more evidence is gained on impact of specific practices, it should also help further shape and recode scientific publishing practices and processes.

A glossary of terms is included for your convenience to provide context and as reference. Also, a list of suggested reading is provided at the end of the book for further reading.

I hope you find this book insightful and inspiring to raise the bar further for scientific publishing. I also hope that it helps stimulate new ideas and suggestions to continue our momentum of bringing more positive transformations and recoding of scientific publishing for our future generations.

Why Publish Scientific Research?

I finished my first book seventy-six years ago. I offered it to every publisher on the English-speaking earth I had ever heard of. Their refusals were unanimous: and it did not get into print until, fifty years later, publishers would publish anything that had my name on it.
– George Bernard Shaw

The publishing hurdles and dilemma that George Bernard Shaw faced a century ago were similar to Albert Einstein's experience and frustrations with scientific publishing as described in the biography *Einstein: His Life and Universe* (authored by Walter Isaacson), which continues to haunt scientific researchers to this day. So the first question that may come to mind is—if it is such a cumbersome process, why bother to publish scientific research? The answer to this question may seem too obvious or somewhat of a cliché. Publication of scientific research allows researchers to share their findings with fellow researchers and the scientific community. It is a fundamental process that serves to increase awareness of new learnings, foster intellectual discussion and debate, and expand the wealth of scientific knowledge and experience across the world, which ultimately helps propel scientific and technological advancements for the betterment of humanity. Scientific or scholarly publishing is the established medium of communication for almost all disciplines—physics, mathematics, statistics, biology, chemistry, medicine, life sciences, technology, and many more. For the purposes of this book, we will focus on publication of research related to advancing health and to science related to treatment, prevention, and curing disease states and conditions. For this, the World Medical Association's (WMA) "Declaration of Helsinki on Ethical Principles for Medical Research Involving Human Subjects," the widely accepted and referenced founding principles for conducting clinical research, considers publication as an ***ethical obligation,*** saying:

> *Researchers have a duty to make publicly available the results of their research on human subjects and are accountable for the completeness and accuracy of their reports.*
>
> *Negative and inconclusive as well as positive results must be published or otherwise made publicly available.*

Along with this obligation, the declaration includes other ethical requirements such as risk minimization, protection of privacy of research subjects, confidentiality of personal information, and informed consent of human subjects or patients participating in clinical trials. Although the obligation to publish is widely recognized and accepted, there are still ongoing concerns about clinical research remaining unpublished.

Here, it's important to better understand what "publication" means and its history. In most dictionaries, the broadest definition of publication is *communication of information to the public.* The word *publication* derives from the Latin word *publicare,* meaning "to make public." It is important to note this underlying meaning of publication…to make public. Historically, information was made public primarily through verbal communication or personal writing. The most widely known examples are ancient scriptures such as the *Vedas,* the *Quran,* and the *Bible.* The invention of the printing press in the fifteenth century revolutionized how we communicated and allowed us to have more systematic and wider distribution of knowledge and information. These advances in communication and standards of living eventually resulted in the emergence of scientific communities such as the Royal Society of London as well as foundational texts such as Newton's *Principia* in the seventeenth century. The Royal Society of London adopted a practical format at the time for exchange of research work—the **periodical journal**—and thus the first scientific journal in English was born in March

1665 as the *Philosophical Transactions of the Royal Society*. Of note, the first scientific journal in any language is believed to be the *Journal des sçavans* published in France in January 1665. The periodical journal was a clever and pragmatic approach to having a scholarly exchange of information, research work, and scientific debate at the time, as there were inherent resource limitations associated with printing on paper and distribution of the printed journal by mail. Over the years and centuries, the number of scientific journals has now proliferated to over 28,000 active scholarly peer-reviewed English-language journals and an additional 6,450 non-English journals as of 2014 (Ware and Mabe, *The STM Report*, 2015). According to research from the University of Ottawa, as of 2009, there have been over fifty million science papers published since 1665. In addition to the overwhelming wealth of scientific information, the scientific community also has to deal with sifting through an increasing number of predatory or fake scientific journals that publish non-peer-reviewed, poor-quality research. Interestingly, the current model of scientific publication fundamentally remains the same—as the periodical journal—over three centuries later, even after the advent and use of technology.

With the wider acceptance of scientific journals as the standardized form of research communication, it also gave rise to ranking of journals based on reputation, value, and prestige. The most highly prestigious life sciences journals that most researchers want to publish in include *Nature, Science, the New England Journal of Medicine* (NEJM), *the Lancet, the Journal of the American Medical Association* (JAMA), *the British Medical Journal* (BMJ), and the like. Getting your research published in any of these journals is highly competitive with acceptance rates as low as 5% to 11% among these journals, while the overall acceptance rate across journals is around 50%. The concept of journal review and consequent acceptance or rejection of a paper can be attributed to the origins of how periodical journals came into existence. There was a finite and limited number of papers that a journal could print in each periodical issue due simply to logistical constraints of printing on paper and mail distribution.

Importantly, journal peer review became and still remains a quintessential foundation for ensuring scientific integrity of a report, and serves as a mechanism to ensure reporting of high-quality research while sifting out poor-quality research that could be misleading. This is extremely and particularly important for medical literature as it can have a direct impact on health care decision-making, clinical practice, and ultimately the health and safety of patients. In its origin, journal review was conducted primarily by journal editors. The current practice and format of review conducted by external peer reviewers (who are not journal editors) can be considered modernization of the peer-review process. While it is vital to maintain a peer-review process for quality control, we continue to see papers rejected by traditional journals for non-quality related reasons such as lack of novelty or the data being uninteresting.

The journals are highly selective in terms of which articles or papers they will accept for publication. The journal editorial and peer-review process is intended to be a service to the scientific community to ensure publication of high-quality research papers. Yet we continue to see instances of papers being rejected outright without going through peer review, despite the fact that the research topic meets the journal's scope and target audience or readership and is well-written with relevant high-quality research. In a 2016 editorial published in *Nature*, Kendall Powell elegantly describes the long and arduous journey of a budding researcher, Danielle Fraser, who experiences frustration with the number of rejections her paper goes through before finally getting published in *PLoS ONE* 23 months after she had first submitted it to *Science*. This sort of practice begs the question of what might be the real driving or motivating factor for journals to reject a paper. While it is difficult to fully understand, some have criticized that the practice of acceptance and rejection of papers allows the journals to maintain their rejection rates, ranking, and prestige. By rejecting more papers and hence having a higher rejection rate, there can be a perceived higher value to getting a paper actually accepted and published in the journal. As mentioned above, outright or immediate rejections prior to peer review can also occur if the journal's editorial office does not consider the research novel or interesting, which are subjective reasons and do not relate to quality of the research. Such journal practices have a significant negative impact by delaying publication of research findings, thereby posing as a disservice to the scientific community and potentially slowing scientific advancement for humanity.

Academia has also been an intricate player in sustaining the current model of scientific publishing in journals. Academia has been known to put pressure on their researchers through "publish or perish" phenomenon. It is well known and established that the tenure and promotion of academic researchers are heavily rewarded and measured based on number of publications and where the researcher publishes his or her research. Research published in high-ranking journals or those with high impact factors can greatly increase the chances of promotion for the researcher. Therefore, researchers are more likely to continue to have their research submitted to and published in traditional journals instead of considering and utilizing alternative methods of making their research public. This practice by academia unfortunately continues to fuel the demand for traditional journals and encourage the journals to carry on with the practice of high rejection rates (which gives perceived higher reputation) and managing journal impact factors. Many have criticized this academic practice and called for a reform in how academic institutions reward researchers for their work. However, academic institutions are not the only ones with such ingrained perceptions and beliefs regarding journal publication. Regulatory health authorities; the pharmaceutical, biotech, and medical device industry; and government and non-government research grant providers hold similar beliefs and differentiate results posted on a public website from journal publication, giving journal publication more credibility and value. If the research and scientific community would like to change the current landscape and model of publishing, then all stakeholders would need to shift their mindset from the traditional journal publishing model to more innovative publishing models that are grounded in the original idea of making research public while accelerating the publication process, maintaining quality through peer review, and elevating the value of the research itself rather than the journal it is published in.

One such movement toward increasing the value of research is seen in the emerging new type of publication metric called the alternative metric, which measures dynamic attention indicators at the article, rather than journal, level (www.altmetric.org—note that this is different from www.altmetric.com, a company that provides the tool to generate Altmetric score). Altmetric score (www.altmetric.com) is an article-level metric that measures the attention a research article or dataset receives online as opposed to measuring its success based on the journal's overall ranking or impact factor. The Altmetric score for each article is calculated based on the level of attention by the volume of mentions relative to each source type. The types of source included in this calculation take into account our current mode of social communication, including but not limited to:

- News, blogs, Twitter, Facebook, Sina Weibo, Wikipedia, policy documents, F1000, YouTube, Reddit/Pinterest, LinkedIn

Intuitively, it makes much more sense to measure the value of research at the article level rather than the article's venue (i.e., the journal) and its overall ranking. It is encouraging to see that many journals have started to report Altmetric scores on the journal's website together with the published article. However, there is still confusion over what the score really means, and it may take time for the academic and scientific communities as a whole to fully grasp, understand, and apply this new system to filter and assess the value of research. The true application and value of alternative article-level metrics is more likely to be appreciated in the absence of journal selection bias, particularly for peer-reviewed articles or datasets that are self-published on an online platform.

In summary, scientific publishing is an ethical obligation as well as a key medium for sharing and exchanging scientific information and opinions. Many key topics related to scientific publishing have been introduced in this chapter; a more detailed discussion follows in coming chapters throughout this book.

Industry vs Non-Industry
Sponsored Research

As medical research continues and technology enables new breakthroughs, there will be a day when malaria and most all major deadly diseases are eradicated on Earth.
— Peter Diamandis

Advances in science and technology to date have enabled tremendous progress in medicine and will continue to provide more breakthrough treatments for diseases and conditions that affect millions of people worldwide. Cancer, which was almost always considered synonymous with "death sentence" in the past, is now managed as chronic disease with longer survival for many, and some are even aiming for a cure! Such breakthroughs and innovations (which are also highlighted in the above quote and vision by Dr. Peter Diamandis, a physician and co-author of several *New York Times* bestseller books) come with significant research and development (R&D) costs, requiring billions of dollars. Health care research is primarily funded by governmental organizations such as the National Institutes of Health (NIH) in the United States or the National Institute for Health Research (NIHR) in the United Kingdom; pharmaceutical, biotech, and medical device companies (collectively referred as "industry"); and other non-profit organizations.

Funding Sources for Biomedical Research

Many rare conditions, which historically were less appealing to the industry for investment of R&D resources, have now become the focus of many companies, resulting in new therapies for patients who in the past had very few or no options. In the United States, a condition is considered a rare disease if it affects fewer than 200,000 people. According to a report issued by PhRMA ("A Decade of Innovation in Rare Diseases: 2005–2015," 2015), there are currently 7,000 known rare diseases and an estimated 350 million people worldwide living with a rare disease, including 30 million Americans (or one in ten). Since the enactment of The Orphan Drug Act in the United States in 1983, which offers certain tax and exclusivity benefits to biopharmaceutical and device companies for developing treatments for rare diseases, the US Food and Drug Administration (FDA) has approved nearly 500 orphan drugs, including more than 230 new orphan drugs, within the last decade alone. In comparison, there were less than ten orphan drugs approved by the FDA prior to the Act. Indeed, the approval and availability of newer therapies has given hope and benefit to patients with rare cancers such as chronic myelogenous leukemia (CML) and chronic lymphocytic leukemia (CLL). Also, those with conditions such as pulmonary arterial hypertension, hereditary angioedema, and cystic fibrosis have benefitted from new therapies over the last decade. However, these treatments cover only 5% of rare diseases, and a lot more work needs to be done. It is encouraging to know that more than 450 orphan drugs are currently in development. Similar advances in research and development have expanded treatment options benefiting much wider groups of patients suffering from major and more common disease states, including cardiovascular, respiratory, and neurological conditions.

The availability of new and innovative treatment options is attributed to financial investments made for biomedical research. The average cost of developing a new medicine has increased significantly. A recent study published by the Tufts Center for the Study of Drug Development (CSDD) reported an average cost of developing one new drug to be $2.6 billion (USD) between the years 2000 and 2010, compared to $1 billion (USD) in the 1990s–2000 and $179 million (USD) in the 1970s. According to the 2015 PhRMA report, PhRMA member companies spent an estimated 17.9% of total sales on R&D in 2014, an increase from 14.4% in 1990. Top medical device companies devote 6% to 12% of revenues toward R&D (Collins, *Market Realist Report*, November 2015).

A recent study focusing on biomedical research funding trends in the United States and internationally, published in *JAMA* in 2015 by H. Moses, et al., reported that collectively (industry and public) global biomedical research funding had increased from $208.8 billion in 2004 to $265 billion in 2011, including combined sources from the United States, Europe, Asia, Canada and Australia. The United States remains as the world leader for both public and industry funding in 2011. Within the United States, total funding increased from $109.7 billion in 2004 to $116.5 billion in 2012. Industry support, including pharmaceutical, biotechnology, and medical device companies, accounted for the majority of funding both internationally (61%) as well as within the United States (58%), while the National Institutes of Health (NIH) was the next largest funder (27%) within the United States. The majority of funding from industry was attributed to clinical research, while NIH funding was largely for basic research (i.e., preclinical).

Since 2000, increasing amounts of funds are also being infused by private charitable foundations such as the Bill and Melinda Gates Foundation. More recently, the co-founder of Facebook, Mark Zuckerberg, and his wife, Priscilla Chan, also announced a $3 billion investment over the next decade to help cure, prevent, or manage diseases, through the Chan Zuckerberg Initiative.

Study Sponsor Versus Funder

Financial funding is one aspect of research, and there are the additional responsibilities of initiating, planning, and managing the trial, all of which are encompassed by whoever takes responsibility as the study's 'sponsor.' In this context, it's important to understand what is meant by sponsor of a trial. Often, it is misunderstood as being associated with financial support only. However, trial sponsorship is regulatory terminology, with a specific definition and associated responsibilities.

The European Commission Directive 2001/20/EC defines the sponsor as:

> *An individual, company, institution or organisation which takes responsibility for the initiation, management and/or financing of a clinical trial.*

The US FDA (21 CFR 312.3(b)) defines as thus:

> *Sponsor means a person who takes responsibility for and initiates a clinical investigation. The sponsor may be an individual or pharmaceutical company, governmental agency, academic institution, private organization, or other organization. The sponsor does not actually conduct the investigation unless the sponsor is a sponsor–investigator.*

It's important to understand the definition and responsibilities of sponsor as we look further into compliance with clinical trial registry regulations and publications in the next chapter.

Industry-Sponsored Research

Industry-sponsored research mentioned in this chapter and throughout the book refers to the above definition of sponsorship, whereby the biopharmaceutical or device company takes full responsibility and ownership of the trial or research. The academic researchers who participate in industry-sponsored trials conduct the investigation as investigators.

Collectively, the drug and device industry worldwide plays an important role and is the powerhouse of bringing newer therapies and treatment options to patients. R&D of new drug therapies is a very complex, expensive, and lengthy process, requiring more than ten years on average to go through the entire process from the time the compound is identified to when it receives regulatory approval. The pharmaceutical and biotechnology industry is regulated by the FDA in the United States, by the European Medicines

Agency (EMA) in Europe, and by individual country laws and regulations worldwide.

In comparison, the R&D process for medical devices is less complex and much shorter at about 18 to 24 months for a very simple device to more than 30 months for a device that requires FDA 501(k) approval. This could become longer as new regulations emerge and are put in place to address the complexities of new technology in light of patient safety. Class II medical devices, such as shoulder prostheses, require FDA premarket notification [510(k)] and some require provision of clinical data to the FDA, while all Class III devices (e.g., cardiac stents) require FDA premarket approval via a premarket approval application (PMA), which must include clinical data supporting the safety and effectiveness of the device. Furthermore, the FDA issued a draft guidance in July 2016 for use of real-world evidence to support regulatory decision-making for medical devices, thereby highlighting the importance of research and evidence for medical devices. At the time of writing this book, the final guidance was pending. The medical device industry is also regulated by the European Commission for approval and marketing authorization in Europe, as well as individual country laws and regulations worldwide.

In line with the responsibilities associated with study sponsorship, the industry is also expected to take responsibility and accountability for ensuring that the clinical research sponsored by the company is made publicly available through scientific publications and disclosure on public registries and websites. Clinical research transparency and disclosure requirements apply to all industry-sponsored research—pharmaceuticals, biotech, and medical device. Most major biopharmaceutical companies have company policies for clinical trial disclosure and publication that are aligned with regulatory requirements, and outline further commitments related to journal publications, including timing of manuscript submission. The commitments related to the scope of trials, timing of manuscript submissions, etc. generally vary from company to company.

A friend and colleague of mine used to say, "Publications are the bread and butter of a [pharma] company." This is so because no matter how many studies a company may conduct, ultimately, they can only be helpful if the studies get published. Likewise, publications are also utilized to support marketing claims and regulatory approval, and essential for preparation of educational materials for health care professionals and patients.

Investigator Initiated Trials

What are investigator initiated trials (IITs)? These are generally unsolicited clinical studies, where an individual, institution, or organization *other than* the commercial or funding entity takes full responsibility for the initiation, management, and regulatory compliance of a clinical trial. The entity that takes full responsibility for the clinical trial plays the role of both sponsor and investigator. Hence, for IITs, the investigator or institution is the study sponsor and responsible party, not the funding pharmaceutical/biotech/medical device company. In other words, these can be industry-funded, but are not industry-sponsored trials. The company may provide either financial support or funding, drug supply, or both for IITs. IITs can be with or without company drug involvement, and they can be interventional or non-interventional studies, including prospective registries or retrospective chart reviews. IITs can also be funded by private non-profit organizations or government organizations.

IITs are generally conducted to expand the knowledge and understanding of a disease or the use of treatments. Due to nature of these studies as independent from industry involvement, they generally hold more perceived credibility and carry greater weight. It is important that both industry and investigators of IITs respect this independence for publications that are generated from IITs as well.

It is an IIT sponsor's responsibility to ensure trial disclosure on regulatory repositories and to publish the results. Since trial registration and results disclosure for IITs are the responsibility of the IIT investigator rather than the funding company, there may be limited monitoring to check if all IITs are in fact being publicly disclosed. Unless the funding company has a rigorous monitoring program for IITs and their publication, the publication of IITs can also remain un-checked.

Similar to study conduct, IIT investigators need to take full responsibility and maintain independence from the funding company for all aspects of IIT publication development, including compliance with ICMJE publication guidelines. Often, IIT investigators request medical writing support from the funding company for IIT publications after the IIT is completed. If the company proceeds to provide medical writing support after IIT completion, it can be perceived as cherry-picking and should be discouraged. Therefore, it is important that the IIT investigators request the funding for medical writing assistance for publications up front, within the original study budget. The IIT investigator should also take responsibility for directly selecting and hiring the professional medical writer, if needed, without the influence or involvement of the company that funded the IIT or provided the drug supply. Again, this is to ensure maintenance of independence of IITs. Professional medical writers and editors can be found through freelance writing services or the directories of associations such as the American Medical Writers Association (AMWA, www.amwa.org) or European Medical Writers Association (EMWA, www.emwa.org).

Due to the collaborative working relationships that the IIT investigators may have with industry-employed medical associates for other studies, the IIT investigator may occasionally wish to include industry employees as co-authors on IIT publications. However, due to the independent nature and definition of IITs, no industry employee would qualify for authorship of IIT publications, according to the established ICMJE authorship criteria (www. icmje.org), as they would not have provided a substantial contribution to the study. Hence, inclusion of industry employees as co-authors on IIT publications may not be justified. It is essential for companies to set policies that clearly describe the roles and responsibilities of the IIT investigator (third-party sponsor) and the funding company to ensure full independence of IITs and their publication.

Non-Industry-Sponsored Research

Non-industry-sponsored research can be categorized as either government-funded or non-government-funded (e.g., funded by non-profit organizations). Non-industry-sponsored research must also adhere to similar standards and accountability for public disclosure of studies and their results. Each funding provider may have their own policy for publication and disclosure of research that is funded by the organization.

Government funds innovation through basic scientific research, which is then further built upon by industry to eventually bring innovative treatments to patients. According to the US NIH public access policy (https://publicaccess.nih.gov/), the grant recipients are required by law to submit the final accepted version of a peer-reviewed manuscript to PubMed Central, an online public repository:

> *All investigators funded by the NIH submit or have submitted for them to the National Library of Medicine's PubMed Central an electronic version of their final, peer-reviewed manuscripts upon acceptance for publication, to be made publicly available no later than 12 months after the official date of publication: Provided, that the NIH shall implement the public access policy in a manner consistent with copyright law.*

Recently, in September 2016, the NIH issued a complementary policy in agreement with the final ruling issued by the US Department of Health and Human Services (DHHS), requiring academic institutions and organizations to register and post results of all NIH-funded clinical trials to ClinicalTrials.gov, including those that are not part of the regulations such as Phase I studies, trials of behavioral interventions, or non-FDA regulated products. The UK NIHR also has similar policies for clinical trial disclosure and publication (www.nihr.ac.uk/policy-and-standards/publications-policy.htm).

The Bill and Melinda Gates Foundation, noted as one of the largest private non-profit foundations to provide funding for biomedical research in US (Moses et al., *JAMA*, 2015), recently released one of the strongest open access policy effective January 1, 2015, and stated: (http://www.gatesfoundation.org/How-We-Work/General-Information/Open-Access-Policy):

> *We have adopted an open access policy that enables the unrestricted access and reuse of all peer-reviewed published research funded, in whole or in part, by the foundation, including any underlying data sets.*
>
> *All publications shall be published under the Creative Commons Attribution 4.0 Generic License (CC BY 4.0) or an*

equivalent license. This will permit all users of the publication to copy and redistribute the material in any medium or format and transform and build upon the material, including for any purpose (including commercial) without further permission or fees being required.

Open access to the publication can be on the journal's website or PubMed Central (immediately upon journal publication), and the Gates Foundation will pay each article's processing and open access fees. Based on this, researchers will need to ensure that publications based on research funded by the Gates Foundation are published in journals that do not withhold open access of publication on the basis of embargo or copyrights. This is a big move given that many major journal publishers impose an embargo of about six months or a year before allowing authors to archive the published articles on their own or their institution's website. At the outset, this will likely create restrictions for grant recipients barring publication in journals that hold embargo policies, but hopefully, the publishers start taking a look at such embargoes and restrictions and update their policies to address the mandates placed by research funders.

The second hurdle to overcome is related to copyright, as most traditional journals take over the copyright from the authors when they accept the manuscript for publication. While it is important to protect a work's copyright and attribute credit to the creator of the work, it seems a bit strange that the rights of the work that an author has created (figures, graphs, tables, etc.) for the manuscript transfer over to the journal publisher, who technically was not the original creator of the work. Researchers have probably willingly given up copyrights as they felt they had no choice but to do so in order to get their paper published.

The transfer of copyright helps generate a revenue source for journal publishers through reprints purchased in bulk by biopharmaceutical and device companies for distribution (which is considered commercial use). In addition, many traditional journal publishers charge fees ranging from under $100 to several hundred dollars for a single use of each graph, figure, or table from published articles, in particular when it's for commercial use. According to Richard Van Noorden (*Nature*, 2013), data from a report of firm (Outsell of Burlingame, California) suggests that the publishing industry generated $9.4 billion in revenue in 2011 with ...nated profit margins of 20 to 30%. The traditional journal publishers are likely to be resistant to changing their copyright policies in order to sustain their revenue source. This should, however, change gradually over time given the success of open access publishers with journals such as *PLoS ONE* and others that have changed the landscape and publish the articles under CC BY license (see glossary of terms for further details). However, not all open access journals are created equal. The copyright licenses can vary from journal to journal; hence researchers need to be cognizant of the journal's copyright policy when selecting one for publication.

The UK Wellcome Trust, a biomedical research charity that has provided grants to over 14,000 recipients from over 70 countries, has a slightly different approach for its open access policy (https://wellcome. ac.uk/funding/managing-grant/open-access-policy):

> *We expect our researchers to publish as high-quality, peer-reviewed research articles, monographs and book chapters.*
>
> *…require electronic copies of any research papers that have been accepted for publication in a peer-reviewed journal, and are supported in whole or in part by Wellcome Trust funding, to be made available through PubMed Central (PMC) and Europe PMC as soon as possible and in any event within six months of the journal publisher's official date of final publication (similarly, monographs and book chapters must be made available through PMC Bookshelf and Europe PMC with a maximum embargo of six months).*
>
> *…encourage—and where it pays an open access fee, require—authors and publishers to licence research papers using the Creative Commons Attribution licence (CC-BY) so they may be freely copied and re-used (for example, for text- and data-mining purposes or creating a translation), provided that such uses are fully attributed (CC-BY is also the preferred licence for monographs and book chapters).*

In contrast to the Gates Foundation and NIH policy, the Wellcome Trust policy appears to be more flexible, allowing the researchers to publish the research in the form of a peer-reviewed journal article, book chapter, or monograph (these are generally published by University Press). It also differs from the Gates Foundation policy with regard to timing of open access and provides a six-month allowance, while the copyright requirements appear to be similar.

In summary, although we will not be able to cover the policies of all non-industry research funders, the key takeaway is that researchers need to pay close attention to the individual funder's publication policy—one size does not fit all. In addition, with the increased spotlight on transparency and open access, scientific publishing will continue to evolve as major research funders establish more open access policies. This is an area where the biopharmaceutical and device industries can adopt as best practice to further improve on transparency, and commit to publishing industry-sponsored research exclusively in an open access forum.

Thus far, we have gotten an understanding of the different categories of research according to funding source and study sponsorship. In the next chapter, we will review the regulations, and compliance with those regulations as well as publications for industry-sponsored and non-industry-sponsored research.

Clinical Trial Disclosure and Publication in the Era of Transparency

A lack of transparency results in distrust and a deep sense of insecurity.
– His Holiness the Dalai Lama

With increasing demands from regulators and funders, it is likely that the first thing many researchers think of is "What am I required to publish or disclose?" Most are inclined to comply with anything that has legal obligations. So let's first take a look at what is required by law. While there is no law that mandates publishing medical research in peer-reviewed journals, there are regulations around clinical trial registration and disclosure of results on publicly accessible databases.

Clinical Trial Disclosure Regulations and Requirements

Clinical trial transparency through registration and results posting increases awareness among the medical community and researchers of planned, ongoing, and completed clinical trials; helps prevent unnecessary duplication of research effort; and could help identify potential collaborators for research. Such transparency also helps build trust by assuring consistency between the protocol–results synopsis on the registry and the study methods and data reported in publications for the same study. ClinicalTrials.gov was initially launched as a voluntary registry but was later converted to regulatory requirement. The US Food and Drug Administration Amendments Act of

2007 (FDAAA) mandates registration of interventional clinical trials of any FDA-regulated drug, biologic, or device, regardless of funding source, on ClinicalTrials.gov no later than 21 days after enrollment of the first study participant. Studies that are excluded from this registration requirement are Phase I trials and small feasibility studies of devices. The FDA defines Phase I trials as those involving 20 to 100 healthy volunteers or patients with the disease or condition being investigated. Furthermore, FDAAA also requires that the results of these clinical trials are posted on ClinicalTrials.gov within 12 months of completion of data collection for the pre-specified primary outcome (also referred to as primary completion date). This regulation applies to all clinical trials regardless of funding source, which can be industry, government (such as NIH), academic institutions, or any other funding organization. The final ruling of this regulation, issued in September 2016, now enables the FDA to enforce monetary fines for those who fail to comply. At the same time, the NIH issued a similar, complementary policy requiring registration and results reporting on ClinicalTrials.gov for all NIH-funded trials (including trials that may be out of scope under FDAAA).

Similar requirements also exist from the European Medicines Agency (EMA) for trial registration and results disclosure, at https://eudract. ema.europa.eu/. The EMA requires registration and results posting for all interventional clinical trials (including phase I through IV) of marketed or investigational drugs, which involve at least one participating site in the European Union. Results of studies involving adult patients are required to be posted within 12 months after study completion, while those involving pediatric patients must be posted within six months after study completion. The agency has also more recently declared further transparency to publicly disclose regulatory approval applications and related documents such as study protocols, clinical study reports, etc., regardless of whether the drug has been granted approval or not. To further support transparency and enhance compliance, the ICMJE guidelines, which are followed by the ICMJE member journals and adopted by many non-member journals, require that the clinical trials be registered on ClinicalTrials.gov or similar registry in order to be considered for publication in their journal. This requirement in part helps to detect and prevent publication of selective results. Furthermore, the industry associations such as PhRMA, EFPIA,

JPMA, and IFPMA have also declared their support for transparency, and the Joint Position (http://www.efpia.eu/uploads/Modules/Documents/20100610 joint position publication 10jun2010.pdf) goes further to recommend that companies should also publish the results of Phase III trials in scientific peer-reviewed literature in addition to posting the results on public websites.

Based on these regulations and guidelines, the industry and the research community have embraced and recognized the importance of transparency for clinical trials. Companies have dedicated resources committed to ensuring that clinical trials, as required by regulators, are registered at the start of the study along with results being posted after trial completion.

Compliance with Clinical Trial Registration and Results Disclosure

Many studies have been published over the years assessing compliance with the FDAAA regulations for clinical trial disclosure. Despite the awareness of and recognition for clinical research transparency among all stakeholders (regulators, academic researchers, the industry, publishers), we continue to see reports of lack of compliance with clinical trial registration and results disclosure. Although the industry has been criticized in large part for under-publishing clinical research, studies have in fact shown that the lack of compliance is not confined to industry-sponsored studies alone.

Anderson et al. (*NEJM*, 2015) reported that of the 13,327 trials that were completed or terminated between 2008 and mid-2012, 13% had reported results on ClinicalTrials.gov within 12 months after trial completion and a total of 38% reported during the five years following study completion. Although a majority of the trials were industry-funded (66%), 14% of the observed studies were funded by the National Institutes of Health (NIH) and the remaining 20% by other government or academic institutions. Interestingly, the 12-month reporting rates were very low regardless of the funding source: 17% for industry-funded; 8% for NIH-funded; and 6% for trials funded by other government or academic institutions. The reporting rates improved across all funding source at five years post study completion, to 41%, 39%, and 28%, respectively. In a multivariate analysis, the authors reported that industry-funded trials were more likely to report results in a timely fashion, while trials funded by other government or academic institutions were less likely to report on time. Median time to reporting was 16 months for industry-funded, 23 months for NIH-funded, and 21 months for other government-funded or academic institution trials.

Another study by Miller et al. (*BMJ Open*, 2015) reported a 67%-71% rate of results posting according to FDAAA requirements for 318 industry-sponsored trials specifically involving 15 new drugs that the FDA had approved in 2012 for ten large biopharmaceutical companies. Furthermore, out of 807 industry-sponsored trials for 53 medicines approved by EMA between 2009 and 2011, results were disclosed within 12 months for 77% of the trials (Rawal and Deane, *CMRO*, 2014). When extending the reporting time beyond 12 months, the reporting rate increased to almost 90%. This study included multiple publicly available sources for results reporting, including ClinicalTrials.gov, the EudraCT website, company websites, and the IFPMA clinical trials portal.

Most recently, Chen et al. (*BMJ*, 2016) reported that the results reporting rate on ClinicalTrials.gov was 13% after 24 months post study completion for 4,347 studies with primary completion date between October 2007 and September 2010 that were registered with academic institutions as the responsible party. The study included trials from prominent US academic institutions such as Brigham and Women's Hospital, The Cleveland Clinic, Columbia University, Cornell University, Dana-Farber Cancer Institute, Duke University, Johns Hopkins University, Massachusetts General Hospital, Mayo Clinic, MD Anderson Cancer Center, Memorial Sloan-Kettering Cancer Center, Stanford University, University of California (several locations), Yale University, and 37 more across the United States. Across the 51 institutions, results reporting at 24 months post study completion ranged from 1.6% to 41% of trials. With a longer follow-up period post study completion, the overall reporting rate remained low, ranging from 4% to 55%. Median time to results reporting after study completion was 14 to 47 months across institutions.

Such alarming reports raise an important question—why do we see non-compliance for something that is legally required? Here are some insights on potential reasons:

- Until September 2016, due to the lack of a final regulation, the FDA was unable to enforce the $10,000 per day penalty for non-compliance with trial registration and results posting. Therefore, the compliance data observed and reported to date reflect the behavior of the medical research community in light of less stringent regulation without any enforced penalty or negative consequences for non-compliance. It would be of interest to see if there is an improvement in compliance with this regulation in the future.

- There could be a misconception among academic researchers that the FDAAA requirements only apply to industry-sponsored clinical trials. Lowest reporting rates were observed for trials funded or conducted by academic institutions. Unlike industry-sponsored trials, which can be accounted for through publicly available health authority application documents (such as clinical study reports and other documents submitted to the FDA or EMA), accountability for academic institution-sponsored trials remains elusive. For these trials, there is no baseline repository to confirm if all trials managed by academic institutions have been accounted for through trial registration, results posting, or publication.

- Another possible reason for low reporting rates could be related to cumbersome website interfaces and resource constraints. Preparation of a results summary and posting it on ClinicalTrials. gov can be time and resource intensive, partly due to its outdated interface. Indeed, one company has reported that it can take five to 60 hours to prepare results summaries. Consistent with the higher reporting rates seen with industry-funded trials, biopharmaceutical companies are able to complete the required summaries due to appropriate infrastructure and resources, while smaller organizations, academic institutions, and individual researchers may not be able to allocate enough resources to comply with such requirements. Now, given the final regulation on clinical trial registration and results reporting, academic institutions and researchers will need to ensure adequate allocation of resources to meet the obligations, which will likely come through increases in requested grant budgets to the funders (such as the NIH and other government and private funding organizations). In addition, implementing technological enhancements to the ClinicalTrials. gov website could help to reduce the effort and time required from researchers, which could improve compliance.

• In support of increased transparency, major biopharmaceutical companies have developed their own publicly accessible clinical trial results repositories (see table for a list of clinical trial websites from select biopharmaceutical companies), and there may be trials disclosed on these websites that are not available on ClinicalTrials.gov.

Clinical trial websites for select biopharmaceutical companies

Company	Clinical Trial Website
Amgen	www.amgentrials.com
Eli Lilly & Company	www.lillytrials.com
GlaxoSmithKline	www.gsk-clinicalstudyregister.com
Merck	www.merck.com/clinical-trials/
Novartis	www.novartisclinicaltrials.com
Pfizer	www.pfizer.com/research/research clinical trials/trial results

• Much of the studies assessing compliance to results reporting have focused on a single source, ClinicalTrials.gov. As expected, the study that included multiple publicly available resources (e.g., ClinicalTrials.gov, the EudraCT website, company websites, and other repositories) showed higher reporting rates for industry-sponsored trials (Rawal and Deane, *CMRO*, 2014).

• Definition of primary completion date may vary from company to company based on interpretation. Certain therapeutic areas such as oncology involve trials that run for several years and some even for a decade. It's typical to have first primary data analysis at one or two years from study start while the study continues for another five or ten years. In such cases, some may consider the primary completion date to be at the one or two-year time point, while others may consider the final primary completion date to be at the five or ten-year time point. This variation can have a significant impact on reporting rates depending on when they are measured for compliance.

- Miller et al. (*BMJ Open*, 2015) also noted several factors for potential non-compliance. Ambiguity and lack of clarity on type of applicable trials, timing of reporting, etc. can create a wide variation in interpretation of FDAAA and therefore in its implementation. The authors learned of this during their discussion and input from the companies that were involved in their research. Mergers, acquisitions, and collaborations may also create confusion with regard to who is responsible for completing the registration and results posting. Hence, clarity of the requirements could improve consistency with implementation and thus compliance.

Compliance with Publishing Clinical Research

Now let's take a look at the publication rate for clinical trials. As mentioned previously, publication in scientific journals is generally not legally required; however, there is an ethical obligation to patients that is enshrined in the World Medical Association's Declaration of Helsinki and elsewhere that all clinical trials should be published. Here again, we see similar trends as those noted for results posting. The *BMJ Open* 2015 (Miller et al.) study mentioned previously also reported publication rates for the 318 industry-sponsored trials. Percentage of trials published across 15 drugs ranged from 16% to 100%. Although the study did not specifically measure time to publication, the authors applied the same cut-off date of at least one year post primary completion date for determining publication status as they did for results posting. It's encouraging to see a publication rate of around 56% for such a short follow-up period given that it can take as much as over two years to publish in a peer-reviewed medical journal. With regard to NIH-funded trials, Ross et al. (*BMJ*, 2011) reported that 46% of 635 trials were published in MEDLINE-indexed journals within 30 months of trial completion. Median time to publication was 23 months (with a range of 14 to 36 months) from trial completion. In parallel with the trend seen for results posting, trials funded or conducted by academic institutions were noted to have a much lower overall publication rate of 29% at 24 months post study completion (Chen et al., *BMJ*, 2016). Again, median time to publication was 22 months, similar to what was observed in the other study for NIH-funded trials.

While the journal publication rates are less than desired or expected, it's important to note that publications can be difficult to track from PubMed or any other database if the corresponding clinical trial identifier (e.g., NCT number) is not published within the article. Inclusion of such trial identifier is currently not mandatory nor standard practice across all journals, hence there may be trials that are published but remain unidentified. Over the years, it appears that reporting rates (registries and publications) are higher for industry-sponsored trials, while those for studies with academic institutions as the responsible party appear to be low. I am surprised to see this trend since, intuitively, I had expected the publication rate for academia funded or conducted trials to be much higher given that publications are at the heart of academic research. Additionally, academicians rely on publications for job promotions and tenure. Nonetheless, lack of 100% compliance appears to be a systemic problem within the medical research community, regardless of funding source.

One area that is not discussed or mentioned in any of the above papers that investigated publication compliance is congress publications. Congress publications include abstracts, posters, and oral presentations. Abstracts undergo pseudo peer review by the scientific committee of the congress prior to its acceptance or rejection for presentation. Unlike journal peer review, there is no feedback provided by the congress committee to the authors for revising the abstract prior to acceptance and publication. Hence, given the less rigorous review process, abstracts in general are not considered peer-reviewed publication. Congress abstracts are published either in print or online. Unlike journal articles, they are not MEDLINE-indexed and hence not readily accessible in a centralized free public repository (PubMed) but are scattered among websites or journal supplements of the respective scientific congresses.

Abstracts of certain congresses are searchable and accessible on Embase and Scopus, both of which are subscription-based databases owned by Elsevier (the Netherlands). Scientific posters and oral presentations do not undergo any congress review prior to presentation and may be available online for certain congresses. Nonetheless, the results of clinical trials published as congress publications are publicly available in many instances. Congress publications are much faster to get accepted and published, and are often the first form of publication as soon as the study is completed. Due to the extremely lengthy journal process with multiple rounds of rejections and revisions, scientists often complete congress publication but may give up on journal publication. One of the largest studies, involving 1,075 abstracts presented at a major oncology congress (the American Society of Clinical Oncology), reported that, overall, 39% of abstracts remain unpublished as journal articles four to six years after appearing as abstracts (Massey et al., *Oncologist,* 2016). Moreover, if congress publications are included in the equation, then collectively, the publication rate of clinical trials is likely to be much higher.

Publication Bias

Publication of both positive and negative trials is essential to fully understanding the body of scientific evidence. It is imperative that practicing physicians, allied health care professionals, and patients are made aware of both positive and negative outcomes as this allows the medical community and patients to make informed decisions and treatment choices. Non-publication of negative trials results in reporting bias, which can lead to overestimation or underestimation of the true effect of treatment intervention. Furthermore, publication of negative trials can help save time and resources by avoiding repetition of studies that have shown negative results. Studies have shown that there are more studies with positive outcomes in the literature than those with negative outcomes (Dwan et al., *PLoS ONE,* 2008; Dwan et al., *PLoS ONE,* 2013). Such findings helped raise awareness of the issue of publication bias. Within the last five years, there has been an increasing call for publication of negative trials that has led to the emergence of a number of new journals dedicated to publishing negative results.

Along with a selective study publication bias for positive results, studies have also shown outcome reporting bias and citation bias. In the latest update on systematic review of the empirical evidence of publication and outcome reporting bias conducted by the Reporting Bias Group, the authors reported that 40% to 62% of the studies had at least one primary outcome that was "changed, introduced, or omitted" in the publication when compared to trial protocol (Dwan et al., *PLoS ONE*, 2013). The paper did not clarify whether the discrepancy was explained by the authors in the original study publication. In certain instances, the protocol may have undergone required changes, hence it's important for study investigators to explicitly outline any reasons for discrepancies in the study publication should there be any changes in the analyses compared to the study protocol or if the publication is reporting *post-hoc* analyses. Another aspect of reporting bias has been attributed to inadequate reporting of adverse events or safety data. Inadequate adverse event reporting can lead to misperceptions of the safety of treatment intervention, and thereby pose harm to patient safety. To address this, under the MPIP initiative, joint recommendations by the pharmaceutical industry and journal editors were recently published providing detailed guidance on appropriate reporting methodology (including severity, frequency and timing) along with specific examples as illustrations (Lineberry et al., *BMJ*, 2016). Although these guidelines have been developed for industry-sponsored clinical research, its recommendations can be adopted for any clinical research regardless of funding source.

Furthermore, a recent cross-sectional analysis of thrombolytics for acute ischemic stroke reported that positive trials were cited three times more often than neutral trials and about 10 times more often than negative trials suggesting citation bias (Misemer et al., *Trials*, 2016). Although there were several limitations to this study, it is noteworthy that such citation bias in the literature can lead to distorted views or perception of treatment effect that can influence decision making by health care policy makers and health care professionals.

Clinical trials can remain unpublished in medical journals for several reasons: lack of study publication write-up or inability to get journal acceptance for publication. Regarding lack of study publication write-up, the study investigators and industry sponsors (in the case of industry-sponsored trials) need to take responsibility to ensure publication drafts are prepared for journal submission. In instances where a publication draft is prepared and submitted to a journal, the increased awareness and call for publishing negative study results has helped improve chances of journal acceptance, however, journal rejection still poses one of the challenges. Although many journals promote and support submission of negative results to their journal, the chances of acceptance are generally much higher for high-profile studies that are much anticipated by the medical community, such as pivotal Phase II or III trials. Hence, papers that report negative results from non-pivotal trials often undergo multiple journal rejections before they are lucky enough to be accepted by a journal in the third or fourth or later journal. Often authors give up pursuing submission to another journal after two or three rejections, since each submission to a new journal requires additional time and effort in reformatting to meet journal specifications, in which case it's a loss of authors' time and effort that were spent to develop the draft publication.

The other challenge is the unwillingness of study investigators to submit manuscripts to low-tier specialty journals. More recently, several new journals have cropped up specifically dedicated to publishing negative studies. It's questionable if scientists and researchers read these journals, while some are not even MEDLINE or PubMed indexed. These journals act as placeholders to provide a home for articles that no other journal will accept. Furthermore, the push for publishing all clinical research in journals, coupled with availability of digital technology, may have contributed to the emergence of predatory or fake journals, that have taken advantage of the situation for monetary gains. Lastly, negative trials may get published in non-English journals, which may not be accessible or included in literature reviews focused on English language journals.

When assessing compliance with publication, almost all data is based on whether study results have been published in medical journals. PubMed is the most widely used source for searching and identifying journal articles since it's free. So, the conclusion that negative results remain underreported is specifically for publication in journals, which is only one part of the vast online public domain. On the other hand, if we look at the full Internet landscape, the negative study results may already be publicly available as part of the results posting on company websites or regulatory repositories such as ClinicalTrials.gov and the EudraCT website.

What is the usefulness of having journal publication versus results posting on repositories for negative studies? In comparison to results posting, journal publications provide additional "discussion" section which expresses the opinions of study investigators on the interpretation of study results as well as provides discussion on the relevance of the research findings to previously published literature and to clinical practice. This would be a big void for the studies that are only reported on company websites and registries. In addition, journal publications undergo peer review, while results summaries on regulatory repositories do not. Nonetheless, even though results summaries do not include investigator opinion, implications of the findings, and peer review, the study results or data are what they are. If one is interested in knowing what the findings were from a given trial, the results summaries can provide the answer. For non-clinical research (e.g., preclinical research), the importance of publishing negative findings in scientific journals is much more elevated as this is the only avenue for sharing or communicating research findings. There is no standardized, central repository for non-clinical research and its findings.

One of the approaches that could improve visibility of all clinical trials, including both those with positive and negative results, is the availability of a single portal that pulls and organizes publication information from existing publicly available indexed publication database as well as company websites, regulatory repositories, and other public repositories. One such platform, OpenTrials (www.opentrials.net), was recently launched in October 2016, however, it does not allow users to filter or easily identify clinical trials that have positive, negative, or neutral results. An alternative option would be to publish the negative results in open access, peer-reviewed journals or publication platforms that have greater chances of manuscript acceptance to minimize the time and effort needed in shopping around for a journal to accept and publish. These are potential solutions to either leverage the information already in the public domain and utilizing existing tools and resources.

So far we have reviewed types of biomedical research funding, regulations, publication requirements, and compliance with these regulations and requirements. In the next chapter, we will have a look at the role of publication professionals during the publication process.

The Role of Publication Professionals

Writers aren't exactly people...they're a whole
bunch of people trying to be one person.
– F. Scott Fitzgerald

Publication professionals include a wide range of expertise—medical writers, publication managers, medical editors, and graphic designers. Each expertise can play a vital role during the publication development process.

Professional Medical Writers

Professional medical writers play an important role in facilitating publication development while working closely with and under the direction of study investigators or authors of publication. Although the focus of our discussion is the role of medical writers in scientific publications, it is worth noting that medical writers are also frequently involved in writing regulatory documents such as study protocol, clinical study reports, and other health authority submission documents. Scientific or medical writing is an art and a science. Professional medical writers are individuals who possess both qualities—the unique writing talent (the art) **and** the scientific or medical knowledge and expertise (the science) to be able to write and communicate the results of scientific research in an effective way. Professional medical writers have strong scientific backgrounds and hold high-level academic degrees—PhDs, pharmacists, nurses, and physicians. This is the key factor—given their scientific knowledge, medical writers are able to speak and understand the researchers' scientific language and hence can transcribe it into publication through their writing talent. Study investigators can have the greatest scientific knowledge and expertise but may lack the writing skills to effectively communicate their research. In addition, the most common language in which research is presented or published is English, while the majority of the world is native non-English speaking and therefore needs, or in many cases requires, writing assistance. Journals are likely to reject a paper outright due to poor writing quality.

Two of the four ICMJE authorship criteria related to review and approval of a publication draft requires someone to circulate the draft to all authors (which can often range from a few to more than ten individuals) and collect and consolidate their feedback to generate a revised version. The revised version then goes back to all authors again to ensure that everyone agrees with the changes. This cycle of revisions and draft review typically repeats several times before all authors agree and approve of the content for submission to the congress or journal. This is the tedious and laborious work that academic researchers or clinical investigators do not have the time to do, and the medical writer plays an integral role in facilitating the investigators with the publication draft preparation, review, and approval process.

A recent cross-sectional study (Gattrell et al., *BMJ Open*, 2016) evaluated 233 journal articles reporting randomized controlled trials in 74 BioMed Central journals with articles published from 2000 to 2014. The study showed that articles with declared professional medical writing support had better writing quality than those without medical writing support. Articles with declared medical writing support (n=110) were associated with improved completeness of reporting according to CONSORT checklist, and they were more likely to have acceptable written English according to journal peer reviewers than those without writing support (n=123). Interestingly, the median time to acceptance was 31 days longer for articles with medical writing support. The authors attributed this delay to additional time taken for peer review and for the authors' response to reviewer comments. Another study (Jacobs, *Write Stuff*, 2010) that evaluated 241 articles from a single journal also reported more compliance with CONSORT guidelines in articles involving professional medical writing support than those without.

Although the GPP guidelines recommend disclosure of medical writing support in publications when such assistance is provided, it is a self-reported disclosure and relies on the writers and authors to disclose. Given this, one of the limitations of conducting such investigations on the impact of medical writing support is the accuracy of identifying articles that did not involve medical writing assistance. Despite limited evidence, the improved quality of reporting with medical writing assistance is noteworthy.

Furthermore, academic researchers and authors have acknowledged the value of professional medical writing assistance. In a survey involving 76 academic/clinician authors (Marchington and Burd, *CMRO*, 2014), 83% of authors felt it was acceptable to receive professional medical writing assistance and 84% valued the assistance provided, specifically the assistance related to editing and journal styling, conformity to reporting guidelines such as CONSORT, and the manuscript submission process. Another survey of 415 authors also noted a high proportion of respondents (88%) who considered professional medical writing support to be value added (Camby et al., *Trials*, 2014).

Since the enactment of the US Patient Protection and Affordable Care Act, some (not all) biopharmaceutical companies have reported publication-related medical writing support fee as an indirect transfer of value to the non-industry authors of publications. Some companies may also be voluntarily reporting publication-related medical writing support fees as part of the EFPIA Code on Disclosure of Transfers of Value from Pharmaceutical Companies to Health care Professionals (HCP) and Health care Organizations. Based on this HCP transparency reporting by biopharmaceutical companies, some academic authors may choose to decline medical writing assistance for publications related to industry-sponsored research.

Based on the extent of involvement of professional medical writers in research-related publication development, in a majority of cases, the writers would not qualify to be listed as an author according to ICMJE authorship criteria. The ICMJE authorship guidelines are the industry gold standard, and they state that all authors must meet all four criteria in order to be listed as an author (www.icmje.org):

1. Substantial contributions to the conception or design of the work; or the acquisition, analysis, or interpretation of data for the work; AND

2. Drafting the work or revising it critically for important intellectual content; AND

3. Final approval of the version to be published; AND

4. Agreement to be accountable for all aspects of the work in ensuring that questions related to the accuracy or integrity of any part of the work are appropriately investigated and resolved

For the publications that report research results, in most cases, the professional medical writer is unable to meet the first ICMJE authorship criterion of "Substantial contributions to the conception or design of the work; or the acquisition, analysis, or interpretation of data for the work," as they do not have involvement in the conduct of the study or research, analysis, or interpretation of results. The writer is informed of the study results after the study has been completed and helps collate the information in a publication draft under the direction of the authors. Secondly, although the medical writers may be involved in the writing of the publication, they will not be able to fulfill the last ICMJE authorship criterion to take responsibility and accountability for the research being reported as this responsibility remains with the study investigators. Therefore, in original research publications, for the most part, medical writers will not be eligible to be credited on the author byline according to ICMJE authorship guidelines, however, their contribution should be disclosed in the acknowledgment section.

In contrast, if a professional medical writer provides writing assistance for review articles, it can be argued and justified that the writer should be listed as an author as they could qualify and fulfill all four criteria. It depends on the extent and scope of the writer's involvement in development of the review article. For example, if the writer conducted the literature search, summarized content to write an initial draft, and worked with the academic expert through completion of publication, then he or she may qualify to be an author. However, if the draft was written or substantially revised by the academic expert and the writer simply assisted with administrative tasks such as literature search, copy editing, and formatting, then he or she would not qualify to be an author but should be acknowledged in the publication for their contribution.

Despite the known advantages and significant contributions of medical writers in scientific publishing, the use of medical writing assistance has undergone major criticism and been negatively labeled as ghostwriting in recent years. Ghostwriting is common practice within the general publishing industry where the writer may get *no credit for the writing*, and hence is considered a "ghost." Although controversial and debatable, ghostwriting has also been implicated in ancient scriptures such as the *Torah*, parts of the *Old Testament*, and the *Bible*. The practice remains common and accepted in journalism, movie scriptwriting, fiction, and non-fiction books. There are websites and companies that openly market ghostwriting services for these types of publications.

Scrutiny for ghostwriting (i.e., undisclosed medical writing assistance) in scientific publishing emerged around 2008 from practices described in the preparation of industry-sponsored publications related to rofecoxib (Vioxx), conjugated estrogens/medroxyprogesterone acetate (Prempro), paroxetine (Paxil), etc. Many articles have been published since then on the prevalence of ghostwriting. Criticism of the practice for scientific publications is justifiable since it questions the credibility of the scientific research that is being reported and can be argued to jeopardize patient safety. The media scrutiny was understandable given that, for decades, medical writers hired by the pharmaceutical industry had provided medical writing assistance without receiving any acknowledgment, credit, or disclosure for providing such assistance in the publication. By not disclosing the medical writing assistance and its funding source, it was considered ghostwriting, and moreover, it violated a key public right for transparency since the medical literature is so integral and a key driver of medical practice that affects patients' lives. The good news is that such ghostwriting is no longer accepted in scientific publications. The media scrutiny on ghostwriting of scientific publications was the much-needed wake-up call for study investigators, pharmaceutical companies, journal publishers, medical writers, and anyone involved in scientific publications to change their publication practices and increase transparency. And it has changed dramatically through early engagement of authors prior to initiating publication draft or outline, author agreement to participate in publication development, disclosure of medical writing assistance and its funding in the publication, and disclosure of author involvement in the drafting and development of journal publications—all of which are now part of the industry standard GPP.

Through full disclosure of professional medical writing support, including name of writer, affiliation and funding source of support in the acknowledgment section of the manuscript, there has been a positive change to ensure full transparency. For scientific literature, ghostwriting is described as:

> *Ghost authors participate in the research, data analysis, and/ or writing of a manuscript but are not named or disclosed in the author byline or Acknowledgments.* (Council of Science Editors, White paper on "Promoting Integrity in Scientific Journal Publications," 2012)

Similar definition of ghostwriting was also described by the editors (Laine and Mulrow) of the *Annals of Internal Medicine* in their 2005 article "Exorcising Ghosts and Unwelcome Guests", and it is also part of the position statements issued by the professional societies and organizations such as the Council of Science Editors (as noted in above quote) and ISMPP.

While major pharma and biotech companies have adopted internal policies and procedures for no ghostwriting, it is important that start-up or smaller companies also ensure and adopt principles and practices outlined in GPP and ban the practice of ghostwriting. Academic researchers and investigators collaborating with companies can also take an active role and question the practice of ghostwriting when they see it.

In summary, professional medical writers can have a significant role in providing assistance to authors for publication development. Academic authors seem to recognize and value the assistance, particularly with editorial and administrative support.

Publication Managers

The role of publication managers has grown tremendously, with increasing recognition and responsibilities within the last two decades. This role emerged first at medical communications agencies that provide publication planning and development support to study investigators and industry and was later created and expanded within the industry. More recently, academia has recognized the importance of this role, and some, such as the Duke Clinical Research Institute, have employed their own publication managers to help investigators with management of the publication development process. With the emergence of academic research organizations (ARO) across the United States and Europe, the role of publication managers may continue to further expand, with a unique opportunity to provide publication practice expertise to study investigators and industry sponsors, while being employed and representing the ARO instead of the industry.

On the industry side, the scientific and clinical research is managed and led by the medical organization, and hence it makes logical sense that scientific publications belong and remain within the medical organization in order to maintain scientific integrity and the true essence of scientific publishing. One area that has remained in the gray zone is the group of individuals who conduct and lead health economics and outcomes research (HEOR) and who also publish the HEOR research. Here, the industry has remained variable on which part of the organization HEOR resides in. Due to the "economics" component, historically, this group has been within the marketing organization of a company. However, based on US corporate integrity agreement (CIA) requirements in recent years, many companies have started to shift this group into the medical organization.

A publication manager, who generally resides within the R&D or medical affairs organization of the company, is the *string* that brings the company clinicians or scientists, statisticians, and academic study investigators together by providing expertise in publication planning and development. The role of publication managers at medical communications agencies reside either with account managers, project managers, or medical writers and are an important extension and partner to the industry publication manager. They help manage the time lines and medical writing resources (if medical writing assistance is involved) while ensuring compliance with GPP, ICMJE guidelines, journal and congress requirements, and any other legal or regulatory requirements pertaining to publications.

Bhakti Kshatriya, PharmD

Other Publication Professionals

Aside from these two types of roles, there are various other professionals who may be involved in providing support for publication development, including graphic design, copy editing, proof reading, etc. The extent of support depends on the nature of assistance needed by the authors to complete the publication draft. For example, most often, the journals require figures and graphs included in the manuscript to be provided in a certain high-resolution format (e.g., jpeg or something similar). Authors who may not be proficient with digital software will likely need assistance from graphic designers to professionally draw the figure or graph to make it available in this high-resolution format. Graphic designers can also assist with scientific poster layout according to congress specifications.

Likewise, medical editors, who play a distinct role from medical writer, will often provide necessary copy editing and proof reading support to ensure consistency, flow, and editorial integrity of the publication. Often the editors who are tasked to proof read the publication draft against the original source can flag potential plagiarism, which is against good publication practice. Nowadays, there are numerous software programs and services available specifically to detect plagiarism, which are extremely efficient and helpful to researchers and journal publishers.

Tips and Pearls for Publication Professionals

Over the last two decades, I have had the privilege of working with many publication professionals within the industry as well as medical communications agencies. During this time, I have seen outstanding publication professionals who are passionate and hold the highest ethical standards for scientific publishing. Below is a distillation of over 20 years of experience expressed as key tips and pearls for publication professionals:

- **Always ask yourself what is the *right thing to do*.** As publication professionals, you might often be faced with challenging and sensitive situations like authorship discussions. What has always helped guide me during my career as a publication professional is asking myself "What is the right thing to do?" Asking this question and making it a routine practice, so much so that it just comes to you naturally and automatically, will allow you to manage and overcome many situations or challenges that arise. If you are able to master this, it will provide a huge advantage over those who simply want to follow and do what is *required* of them rather than *doing what's right*.

- **Put yourself in the author's shoes**. Given the intensity and volume of publications that publication professionals generally manage at a given time, it may feel overwhelming and disconcerting when, suddenly, an author provides negative feedback about a publication draft or the process. From an emotional standpoint, the publication professional may feel he or she is trying to do their best and has had only good intentions while liaising with the author on the publication, so wonders how the author could react so negatively. The best way to handle such a situation is to immediately put yourself in the author's shoes. This will help you understand why an author may be raising a certain issue or concern, which will then help you to identify an appropriate solution.

- **Build collaborative relationships by spending more time with the authors on the phone or in meetings**. Having face-to-face meetings or discussions over the phone are so much more efficient and collaborative than trying to gain input simply via email. As a publication professional, whenever I attended a scientific conference or meeting, I would fill up my free time (that is the time aside from attending the scientific sessions at the meeting) by setting up one-on-one meetings or group meetings with authors, specifically to discuss various publications. Those face-to-face discussions were not only the most productive discussions from a work standpoint, but they also allowed us (authors and myself) to connect at the human level. This has been by far the best, most satisfying, and enjoyable part of my career. It allowed me to have scientifically enriched discussions and learn so much from top medical experts from around the world on a wide range of disease states and therapeutic areas. In return, I have received gratifying remarks from medical experts stating that they learned and gained more knowledge about the publication process while working with me. Such remarks came as a surprise to me since the authors were many years senior to me and had published extensively for years before I even met them. I had assumed that they would have more knowledge than me. However, when asked, the response was largely, "No one prior to you spent the time explaining to us why we are following a particular process or completing a task." Taking the time to explain and being transparent about why authors are being asked to do certain tasks, such as completing an author agreement, etc., is often very much appreciated by the authors and also helps build credibility and trust in the publication expertise provided by the publication professional.

- **Ensure appropriate engagement with investigators and authors.** For industry-sponsored research, it is important to engage with investigators and potential authors early on for input on the data prior to initiating publication development, and throughout the course of publication development for critical input and approval of the draft publication prior to congress or journal submission.

  In contrast, requests from investigators for assistance with IIT publications need to be handled differently. In order to maintain the independence of IIT publications, it's important for publication professionals to refrain from activities that can be perceived as influencing the publication. In the instance of IIT publications, if the company-employed publication manager or medical writer provides a copyediting review to the investigator upon the investigator's request—just out of courtesy and a good working relationship with the investigator—the input that was provided, although done as a courtesy, can be perceived as influencing the content of the publication. Hence, many major pharmaceutical companies have limited the company review and feedback of IIT publications to scientific accuracy (e.g., is the data being presented or reported consistent with the conclusions being drawn? Is relevant data missing based on what is described in the protocol? etc.) and clearance from an intellectual property attorney (depending on the IIT agreement or contract). Therefore, in this instance, it would be more appropriate to provide help by suggesting the IIT investigators hire a copy editor or medical writer on their own (e.g., through the AMWA or EMWA website or other freelance writing services).

- **Perception of being commercially driven.** Even though publication managers and medical writers are employed and represent the medical affairs or clinical development division within the company, the external stakeholders will likely not differentiate between medical and marketing, and any feedback provided by a company employee on a publication draft can carry the risk of being perceived as commercially driven. This can occur even if companies ensure that all activities related to publications reside within the medical organization (including budget and human resources), without involvement from associates within the marketing organization.

- **Be nimble and flexible with the changing regulations and publication environment.** Over the years, I have seen publication professionals who are extremely flexible and embraced change, while others frowned upon the changing environment and the constant influx of requirements and regulations. To this, one key advice is to remember that *change is the only constant in life.* Those who are more flexible and nimble and accept the changes are more likely to succeed and continue to grow, while those who resist are likely to produce mediocre results and find themselves going through uphill challenges with minimal to no job satisfaction.

The Making of Scientific Publications

To get to know, to discover, to publish — this is the destiny of a scientist.
— François Arago

The willingness of *doing the right thing* as opposed to *doing what is required* is at the heart of the overall publication process. This underlying principle is crucial and relevant to all stakeholders involved in publication development process: industry, academic/clinical researchers, and publication professionals. The focus of patients can be found in every biopharmaceutical company and health care institution's mission statement, but it can get lost with day-to-day activities. It's not enough to say that the company's mission is better patient care, but to actually *walk the talk* and reflect that the patients are in fact at the center of everything that the health care companies do—it's important to ingrain this mindset into every company employee and within the lifeline and soul of the company's culture.

In this chapter, we will take a deeper dive into the role of the main or key character in the publication development process—the *author*. The role of the author will be highlighted while unfolding the process of publication development.

As a benchmark, I have chosen to describe the publication development process for an industry-sponsored publication, as it has an established, standardized process based on GPP guidelines and also has the most complexity, requiring collaboration and coordination between industry and academic/clinician investigators. Publications developed from non-industry-funded research may undergo a similar process without the involvement of the industry or the funding organization.

Key Highlights of GPP3 Guidelines for Publication of Company-Sponsored Research

The GPP3, published in the *Annals of Internal Medicine* in 2015, builds on the previous versions of the guidelines and is the widely accepted industry standard for principles of good publication practice for industry-sponsored research. It summarizes very well not only how industry-sponsored research should be published, but it also reflects how the industry, at large, currently manages industry-sponsored publications. Major biopharmaceutical and device companies, in general, have embraced the principles outlined in GPP3, which is evident from the individual company's publication policies. Moreover, it is difficult to ascertain the extent of adoption of GPP3 by small to mid-size companies who do not have a publicly disclosed publication policy.

A summary highlighting key points from GPP3 together with my thoughts and comments on each point is provided here. For more details or further information on GPP3, please refer to the published guidelines in the *Annals of Internal Medicine* (http://annals.org/aim/article/2424869/good-publication-practice-communicating-company-sponsored-medical-research-gpp3).

- **Formation of publication steering committee (PSC)** – A PSC is comprised of academic study investigators, and company medical and publication associates. A PSC gives more accountability and ownership of publishing decisions to study investigators and enhances collaboration with the sponsoring company. Thus, avoiding the notion of sponsoring companies being sole decision-makers for study publications.

- **Authorship**
  - Follow ICMJE authorship criteria – These criteria, which are developed by a non-industry organization, are widely accepted and utilized by the publication enterprise.

44

o   Identify and engage authors up front before starting publication draft – this negates the idea of guest authorship.

o   Complete written authorship agreements prior to publication initiation.

o   Payment or remuneration to authors – GPP3 acknowledges that the sponsor companies may reimburse ancillary expenses that the authors may incur, such as travel to congress for poster or oral presentation, or for publication assistance (e.g., medical writing, statistical analysis).

The industry has adopted the principle of no remuneration to authors for publication writing and development to alleviate any perceptions of financial incentive provided to authors, thereby influencing publication content. The academic researchers and investigators have generally accepted this, and it has become industry standard. However, this may not be as formalized for contract agencies that conduct health economic and outcomes research (HEOR). The argument is that HEOR research is commissioned and conducted under an agreement with the sponsor company, and as part of the project, the time and effort spent by HEOR vendor authors on publication writing and development should be compensated by the sponsor company.

This is a weak argument, because the clinical trials are also done under an agreement between the sponsor company and participating institution. However, when it comes to authorship of a publication, there is no compensation to the authors for their time and effort in publication development. Here again is the challenge of changing behavior. Some biopharmaceutical companies have implemented no authorship honorarium for HEOR publications as well as ensuring that the contracts with HEOR vendors do not include any remuneration for publication writing, review, or revisions done by HEOR vendor authors.

○ GPP3 also provides practical, more detailed guidance on how to handle authorship conflicts or unique scenarios (e.g., posthumous authorship when an author dies during publication development before a paper is submitted or published in a journal).

- **Professional medical writing assistance** – Assistance from professional medical writers is acceptable if the authors agree or request it, and such writing assistance must be disclosed in the publication along with the writer's name, affiliation, and funding source. This validates professional medical writer involvement with full transparency and negates the idea of ghostwriting.

- **Scope of trials and timing of publication** – GPP3 recommends that results of all clinical trials regardless of outcome, including interventional and non-interventional studies, be made public. Ideally, this should be in the form of peer-reviewed journal publication, whenever possible. GPP3 also recommends that manuscripts be submitted within 12 months (or latest within 18 months) of study completion for trials related to licensed products or after health authority regulatory approval for investigational products.

- **Publication planning** – This is somewhat tied in with the above bullet. In its most simplistic form, publication planning is essentially a way for researchers to ensure that all studies that have been completed or closed are accounted for in publication. Optimal publication planning involves an integrated approach ensuring that study data are disseminated to appropriate and relevant health care professionals (physicians, nurses, pharmacists, etc.), while taking into account their insights and queries, and addressing potential knowledge gaps. Hence, publication planning is not simply completing one paper per study; it helps identify additional analyses and potential secondary publications that enrich the medical community with knowledge of new discoveries and findings. With increasing regulatory and journal data sharing requirements, optimal planning should also consider timing of and integrate activities around publication of regulatory documents (EMA initiative) and journals' protocol and data sharing requirements. Planning should be a collaborative effort involving all relevant stakeholders, including study investigators, study clinicians, statisticians, publication professionals, and other experts (relevant to the data being reported). It helps study teams and investigators plan for publication activities, which is, in fact, quite a lengthy process—we will go into more detail on it in the next section of this chapter.

The Publication Development Process for Company-Sponsored Research — a Closer Look

GPP3 primarily describes the publication principles, and now we will have a closer look at the overall process of preparing a publication from company-sponsored research—GPP3 in practice! To my knowledge, the detailed publication process has been primarily known by a closed community of publication professionals, researchers, and publishers and may not be widely known among non-research oriented health care professionals. While the process described below is a general set of activities, the policies and procedures followed by individual companies can vary. In addition, although the below description is based on industry-sponsored research publications, a similar approach can be taken by researchers for non-industry-sponsored publications as well.

Publication planning and preparation typically involves formation of a publication team comprised of study clinicians, statisticians, study investigators, publication professionals (a publication manager and/or medical writer), and other experts who may provide expert input on data interpretation. Expert input can be solicited for various reasons if a study involves assessment of a particular adverse event or an evaluation, such as pharmacokinetic data, biomarkers, or patient reported outcomes or quality of life assessments. For example, a cardiologist may be consulted for a cardiac toxicity or a clinical pharmacologist for pharmacokinetic assessments. Ideally, the publication team should be formed early on at the time of study initiation.

Phases of Publication Planning for a Single Study

Publication planning for a study can be divided into three phases:

Phase 1: Prior to study initiation
Phase 2: Close to study completion
Phase 3: At study completion

Phase 1: Prior to Study Initiation

Study Steering Committee (SSC): Many companies establish an SSC prior to study initiation, which allows the company to gain input from academic investigators and medical experts on various aspects of the study, including but not limited to study protocol, design, methodology, patient recruitment, statistical analysis planning, and publication planning. In general, all SSCs include academic medical experts; the number of these experts can vary from three to ten, depending on study complexity and/or company policy or practice. Typically, five academic experts would be ideal. An SSC may also include company employed trial physicians and statisticians, depending on company policy and practice. The company publication manager or professional may also be an SSC member specifically for publication discussions. Some company policies allow equal voting rights to all SSC members including company-employed members and academic experts, while some reserve the voting rights to academic experts only, and company-employed members participate as non-voting members. The latter entrusts the decision-making power to the academic experts. Selection of SSC members depends on the relative expertise of the academic clinician, and their interest and availability to participate as an SSC member. Typically, an SSC also has a chair or co-chairs assigned who play an important role in facilitating discussions and gaining alignment among SSC and publication authors due to conflicting views or opinions. An SSC may be established for non-industry-sponsored studies as well.

GPP3 recommends establishing a publication steering committee (PSC) to help with planning, overseeing, and facilitating authorship decisions for publications from a specific study or group of studies. In practice, it is reasonable to include publication responsibilities with an SSC that is established prior to study initiation, and hence it can play the role of PSC as described in GPP3. If an SSC is not in place for a study, then a PSC can be established prior to study completion. This early timing prior to study completion is important to alleviate any potential bias or preferential selection of SSC members before the study results are available. For publications that combine or pool datasets from multiple studies, it is extremely beneficial to establish a separate PSC to help facilitate authorship decisions, and to provide assessment and validity of scientific hypotheses for publication proposals. In this instance, however, the PSC will be formed after the studies have been completed. As stated in GPP3, SSC or PSC membership does not automatically qualify for publication authorship; however, SSC or PSC members may become authors if they meet all ICMJE (or the congress/journal's) authorship criteria.

Share Company's Publication Policy: It is important to inform all study investigators about the company's publication and authorship policy. The most common practice is to present this at an investigators' meeting prior to study initiation. Often, company publication policy related items may also be included in study protocol and/or a clinical trial agreement. Hence, the investigators should be familiar with the publication policy and guidelines that are applicable to a given study or sponsor. One of the items in the publication policy of many companies is—discouraging the study investigators to publish single-site or single-patient data while the multicenter study is ongoing or prior to publication of primary publication reporting combined results from all participating sites. This is in line with the recommendations in the Joint Position on the Publication of Clinical Trial Results in the Scientific Literature (2010; http://www.efpia.eu/uploads/Modules/Documents/20100610_joint_position_publication_10jun2010.pdf). The reason for discouraging such practice is not to withhold information, but rather to ensure that the first publication from a study is based on the full dataset, including data from all sites, according to study protocol, which is designed and statistically powered to have a sufficient sample size to test the hypothesis. Data from a single patient or single site will not be robust enough to draw scientific conclusions and may even be misleading if these are published prior to primary publication of full study results. However, it may be more appropriate to publish such single patient or site data after the primary publication is published, as it will allow the investigators to present, discuss, and publish the single-site data in the context of full study results. Therefore, it is the pre-primary publication timing of such partial data that is being discouraged to study investigators.

Establish and Share Authorship Selection Criteria: This is probably the most sensitive and critical topic to address up front with all study investigators. By sharing this prior to study initiation, it ensures transparency and helps build trust with the study investigators. This is also recommended in the GPP3 and the five-step authorship framework published by the Medical Publishing Insights and Practices (MPIP) Initiative (Marušić et al., *BMC Medicine*, 2014). The responsibility lies with both parties—the company to ensure that there is transparency in author selection criteria, and with the study investigators to ensure commitment to meet all four ICMJE criteria. Authorship is not an entitlement; it is to be earned.

Phase 2: Close to Study Completion

Identify Authors: For multicenter studies that can range from less than ten to more than a hundred investigators, it is often impossible to include all investigators on the author byline. Author selection is done based on the first ICMJE criterion: *substantial contributions to the conception or design of the work or the acquisition, analysis, or interpretation of data for the work.* However, it is often debatable what can be considered "substantial." To address this, MPIP have put forth practical considerations for companies and investigators to help determine substantial contribution. MPIP recommends tracking investigator participation and contribution during the course of the study, which can help justify why a particular investigator is eligible or qualified to be considered for authorship. In collaboration with the company study team, the SSC or PSC can play an active role in identifying authors based on agreed upon criteria and substantial contribution. For company-sponsored studies, certain company employees, such as a study physician, study statistician, and other scientists such as those involved with clinical pharmacology, correlative sciences, and health economics and outcomes research, may also qualify under the first ICMJE authorship criterion and should be considered for authorship. Including the sponsoring company's scientific and medical employees as authors is also supported by ICMJE authorship guidelines that state:

> *And all who meet the four criteria should be identified as authors.*
> *The criteria are not intended for use as a means to disqualify colleagues*
> *from authorship who otherwise meet authorship criteria by denying them*
> *the opportunity to meet criterion #s 2 or 3. Therefore, all individuals*
> *who meet the first criterion should have the opportunity to participate*
> *in the review, drafting, and final approval of the manuscript.*

The SSC or PSC should also help determine the order in which author names are to be listed, and all authors must agree to the author list, including the order.

Develop a Publication Plan for Each Study: The publication manager plays an important role for this step. Publication planning should be initiated as early as possible prior to study completion. Based on anticipated data availability, an appropriate target congress and journal can be identified for primary publication, which reports the primary analysis of the study. Subsequent secondary publications based on secondary analyses of the study and additional subset analysis can also be identified early on, even before the study is completed. Ideas for secondary publications can continue to emerge for years after study completion, especially for large, multicenter studies involving multiple disease states. As an example, the EPIC (Evaluation of Patients' Iron Chelation with Exjade) study investigating the efficacy and safety of deferasirox (Exjade) in 1,744 patients with iron overload involving a range of hematologic conditions (including thalassemia, myelodysplastic syndrome, sickle cell anemia, aplastic anemia, and other rare anemias) generated over 13 journal articles and many more congress publications over the course of eight years. This does not constitute duplicate publication, which is against good publication practice. However, large, multicenter studies can provide a wealth of information that can be extracted and published in order to provide better understanding of the disease state as well as the efficacy and safety profile of the drug or intervention in various patient populations. This will become even more evident as more secondary publications are generated following ICMJE's upcoming data sharing requirements and recent data sharing initiative implemented by pharmaceutical companies to provide access to anonymized patient-level data to interested qualified researchers.

The practice of encore congress publication or repeating the presentation of the same research data to reach various audiences is generally acceptable when the researchers intend to reach different health care professionals or a different geographical region. For example, research findings that are presented at a society or congress for medical doctors will not likely reach the pharmacists or nurses who also need to be kept abreast of the latest scientific research. Similarly, although major international medical society meetings or congresses, such as the American Society of Clinical Oncology (ASCO) and the European Society of Medical Oncology (ESMO), attract attendance from oncologists all over the world, many clinicians are often unable to attend these meetings and could benefit from having the data re-presented at a local or national meeting or congress. In such cases, the study results may be presented as encore presentations at congresses specifically for specialty health care professionals, or at local or regional congresses. When preparing encore presentations, the authors can consider the target audience or regional nuances and cater the discussion to provide implications of the research findings relative to the target audience. Again, this should not be confused with duplicate publication which is an unacceptable and unethical practice. According to ICMJE, duplicate publications are defined as (www. icmje.org):

> *Duplicate publication is publication of a paper that overlaps substantially with one already published, without clear, visible reference to the previous publication.*

Publication planning and preparation essentially involves project management for publications. As with any project plan, publication plans are living, dynamic documents and require updates on a regular basis. A number of electronic or online platforms are available and are extremely useful in developing and managing publication plans. Planning of publications helps to ensure all key stakeholders within a publication team are informed of the planned publications and time lines. It also serves as a medium for obtaining alignment and agreement among stakeholders for authorship, target congress and journals, timing of congress and journal submissions, primary versus secondary publications, etc. Publication planning is done together with input from all stakeholders: study clinician, statistician, study investigators, and publication professionals. Although publication professionals can help with planning and project management, all final decisions related to publications, including content, data reporting, target congress and journal selection, are made by the authors of the publication.

Key Elements of a Publication Plan

Study Protocol and Data: Study protocol and data are the key ingredients needed for preparing publications. Publication teams should also become familiar with the statistical analysis plan of a study to understand the primary endpoints, secondary endpoints, and other parameters that are planned to be assessed and analyzed. Knowing this allows the team to better prepare for primary as well as various secondary publications. If teams are aware of multiple studies, publication plans can also include planned publications related to pooled analyses or meta-analysis.

Data Availability Date: This is a key piece of information that will dictate the timing of target congress and journal submission. Statisticians play a key role in estimating and confirming when the data can be expected to be available. Publication plans need to be updated if there are any changes or delays with data availability, which can impact congress and journal submission dates and estimated publication dates.

Other Key Dates: There are several other key dates that should be included or considered in publication planning: congress dates and journal dates. *Congress abstract submission date* is a hard date imposed by the target congress. By hard date, I mean fixed dates. If you miss submitting an abstract by the congress deadline, then you have missed the opportunity to submit to that congress, in which case, the authors may need to wait for the next possible congress of interest. Some congresses also issue a *late breaker date*. This is generally meant for special circumstances such as submission of groundbreaking study data or when study results are not available at the time of the regular abstract submission deadline. Congresses have their own guidelines for late breaker submissions. Some congresses announce an *abstract acceptance date*, which is when to expect to hear from the congress regarding acceptance of an abstract. Possible congress decisions include rejection, accepted for poster presentation, accepted for oral presentations, or publication only (but not to be presented at the congress). Certain congresses can also accept abstracts for special presentations such as plenary presentation, which is considered highly prestigious. It's also important to note the *congress presentation date* as this is also another hard date set by the congress, when the author(s) are expected to present their poster or oral presentation at the congress.

Journal submission date is a soft date as the authors are not bound to any submission dates governed by any journal. However, setting a target journal submission date can serve as an anchor and help the team to plan and complete the various tasks and activities necessary to generate a final version ready for journal submission. As this is a soft date, journal submission dates get shifted quite frequently for various reasons—delay in data availability, delay in preparation of publication draft, additional analyses requested by authors during draft review, delay in authors' response to draft review, multiple rounds of draft review by authors, etc. Companies that have commitments for submission of a primary manuscript to a journal within a specific time after study completion may treat the target submission date for primary manuscript as a hard date. Compliance with the company's publication policy is measured using this date. *Journal publication date* is an estimated date based on the individual journal's review and publication times.

Based on the upcoming ICMJE data sharing requirement, it would be important to also include target date for data sharing for that publication. Teams can plan for data anonymization and other activities accordingly.

Type of Publication: Within a publication plan, each type of publication can be identified and mapped out based on whether it is a congress publication or journal publication. As discussed above, congress publications include abstracts, poster presentations, or oral presentations, while there are many possibilities with journal publications, including primary manuscript, secondary manuscript, case report, letter to the editor, brief editorial or communication, etc. Study investigators may also choose to submit a paper on baseline data or a study methodology paper prior to the primary manuscript, especially for large studies involving multiple disease states or subpopulations. Reporting the study results in the form of a letter to the editor or brief communication is also an option, such as for very small studies or studies that were terminated early with very few patients. Letters to the editor and brief editorials are also common for communicating and expressing opinions on certain topics. Case reports are, as the name suggests, intended for reporting individual patient case presentations.

Author(s): As described above, potential authors can be identified by the SSC or PSC according to author selection criteria in reference to ICMJE criterion #1. This can help the team to track author invitation and acceptance from invited authors to participate in publication development. It can also help to track participation from authors during the publication development process as related to ICMJE criteria #2 and 3. The plan should be updated to include the final author list for publication that matches the authors listed in the submitted publication (i.e., those who met all four ICMJE authorship criteria).

Target Congress and Audience: Target congress is selected based on relevant audience according to study protocol and data availability date. For example, if the data of a leukemia study is planned to be available in June, then an abstract can be planned for submission to the American Society of Hematology (ASH) for that year, as their abstract submission deadline is typically in August. However, if the data will not be available until September, then the earliest possible major hematology congress is the European Hematology Association (EHA) in the following year, which has their congress deadline in either February or March.

Encore presentations can be planned based on additional audiences identified relevant to the study (e.g., oncology nurses, pharmacists, etc.) or regional congresses.

Target Journal: It's important to identify and agree on a target journal up front as this will dictate the format in which the manuscript will be prepared (according to journal specifications and requirements). Target journal is typically selected based on a number of factors—disease state and type of data being reported, journal impact factor (although it's questionable on appropriateness of using this as a journal selection factor), journal review and publication time, readership (audience and geographical) in relation to the study, journal's acceptance/rejection rate, possibility of open access, etc. This is a critical decision as, often, targeting a high impact factor journal is associated with increased chances of rejection (Lee et al., *Medical Journal of Australia*, 2006; Siler et al., *PNAS*, 2015). Although it's encouraging to see a 75% acceptance rate with the first journal in a study involving 923 basic biology research journals (Calcagno et al., *Science*, 2012), general experience with medical journals suggests to expect and be prepared for rejection. It is helpful to identify at least one to two back-up journals ahead of time. In order to expedite the publication process, preformatting the manuscript for a back-up journal can help speed submission to the next journal right away, especially if the journal rejects without peer review. The downside to this is that if the first journal does accept the paper, then it was lost time and effort on reformatting.

Measuring the Success of Publication Planning and Post-Publication Impact

Publication metrics can be divided into two major categories—measuring the success and efficiency of publication plan and measuring the impact of a publication after its published. The success of publication planning can be measured based on productivity (volume of publications), efficiency (timing), quality, and compliance. First, let's discuss some of the metrics that measure the success of publication planning:

- **Productivity:** This is typically a measure of volume of publications such as number of abstracts or manuscripts submitted to congress or journal, number of posters or oral presentations presented at congress, and number of journal articles published. These can be measured according to study, group of studies, therapeutic area, intervention, year of submission or publication, individual author, planned versus actual, etc. Detailed analysis of type of data disseminated with respect to audience reached can also help identify potential knowledge gaps. These knowledge gaps can be useful when considering additional secondary publications or for planning new studies.

- **Efficiency:** This involves timing or timeliness of submission or publication. Some examples of efficiency measure are time to submission from study completion, time to publication after first target submission, or time to publication from study completion. Time to submission from study completion is a controllable measure, meaning publication teams can control how quickly the manuscript is developed. Measuring this metric along with deeper analyses allows teams to identify areas for process improvement in draft preparation, author draft review, availability of data or statistical outputs, etc.

- **Quality:** This can be measured in several ways such as quality of writing (can be an indirect measure of number of rounds of revisions required during draft preparation) and acceptance rate at first target journal. These measures are subjective with many confounding factors that can affect them. For example, rejection at first target journal may not be a direct reflection of the quality of a paper or research, it can also be a result of poor choice of the target journal.

- **Compliance:** This is measured to assess compliance with regulatory requirements and/or company/institutional/funder policies for data disclosure and publication. Examples of this type of measure include proportion of completed trials that have been publicly disclosed or published and timing of disclosure or publication according to requirement or policy.

Now, let's review some of the available metrics to measure the impact of a publication after its published:

- **Journal Impact Factor (JIF):** This is a measure of frequency in which articles within a journal are cited within a specific timeframe. Mathematically, it is calculated based on number of citations received by articles published in that journal during the two preceding years, divided by total number of articles published in that journal during the same time frame. This is a measure of the impact of a journal rather than the impact of an individual article. As it is a mathematical calculation, the impact factor can be manipulated by increasing the number of times the articles in a journal are cited. Despite its many flaws and criticisms, it continues to remain the most widely recognized and used measure for publication impact, perceived quality of research, and authors' achievements.

- **Eigenfactor Score:** This is another measure that assesses overall impact of a journal. It takes into account the number of incoming citations for the articles in a journal and the score weighs heavily on citations that are made at high-ranking journals versus those made in low-ranking journals. As with JIF, since this a mathematical calculation, it can be manipulated (e.g., the score can double if the journal size doubles by increase in the number of articles published in a year). Hence, this has similar flaws and limitations as the traditional JIF. Although this is primarily used as a measure for a journal's impact, Eigenfactor scores have been developed for author-level impact as well.

- **Citation Analysis:** This measures the impact of an article based on the number of times an article is mentioned by other authors in their work or publications. Citation analysis can be done using various sources, including Web of Science, Scopus, Google Scholar, and other databases.

- **H-index:** This is a measure to quantify an individual researcher's scientific research output. The index is based on the researcher's most cited papers and the number of citations received for those articles in other's publications. H-index for a researcher can be obtained through various sources, including Web of Science, Scopus, and Google Scholar, and other databases.

- **Altmetrics:** This is a relatively new quantitative measure of attention that a publication receives through social media, citations, and article downloads. Many journals and publication repositories provide altmetrics on their websites. Altmetric data can be obtained through various sources, including Altmetric.com, PLOS, ImpactStory.org, CitedIn.org, and Plum Analytics.

In summary, along with publication planning, it is beneficial to consider how the success of the plan and ultimately the impact of publications will be measured. Measuring such metrics can provide meaningful insights into how well the publication process is working, identify areas of improvement, identify gaps with regards to data dissemination, identify trends in journal response (e.g., acceptance, rejections, actual review and publication times, etc.), and measure compliance with policies and regulations.

Phase 3: At Study Completion

Author Invitation and Agreement: After study completion, publication managers contact potential authors, who are identified together with the SSC, based on agreed upon criteria and contribution. Potential authors are invited in writing via email or publication management system to confirm their participation as an author. It is expected and required to receive acceptance from authors via email or through the publication management tool before engaging the author any further for draft review. This is a key step for process documentation, not only based on GPP3 guidance but also for those companies who are under legal obligations such as the US corporate integrity agreement (CIA), which requires a written author agreement. This should help investigators understand why publication professionals from companies or medical communication agencies chase authors to get their email reply for author invitation and agreement.

At this stage, authors also confirm in writing whether they wish to receive professional medical writing assistance. Authors always have the choice to decline medical writing assistance and may choose to write the publication on their own. Professional medical writing or editorial assistance may include all or some assistance such as writing a draft under the direction of the author(s), draft revisions incorporating feedback from all authors, assistance with graphics and formatting, administrative support including preparation of a submission package, and completing online submission on behalf of the author (upon author's request and approval). Such medical writing or editorial assistance must be disclosed in the publication, including the name and affiliation of the writer, along with funding source. This disclosure is essential and crucial as it enforces transparency and negates the notion of ghostwriting, which is against good publication practice.

In addition, authors should be made aware up front of a company's policy on HCP transparency reporting for the transfer of value in relation to medical writing support for publication. Authors may choose to decline medical writing assistance to avoid being reported.

Author Data Review: Prior to initiating publication writing, study results should be shared with all authors to gain input on data interpretation and conclusions. The most effective and collaborative approach is via teleconference, web conference, or live meeting, if feasible, as it allows the authors to have an open dialogue and discussion about study results. In certain time constrained circumstances, such as congress abstract submission, it may not be feasible to hold a meeting and hence sharing and obtaining input via email may be a practical and appropriate approach. This is a critical activity as it negates the idea or notion of guest authorship, and fosters a more collaborative environment.

All of the above activities we have described are in preparation for publication development, before even starting to write a draft. This prepublication preparation itself takes significant time and resources from the company as well as participating investigators/authors. Now comes the fun part—publication development!

Publication Development and Review Process

Below figure illustrates the development and review process for industry-sponsored research publication.

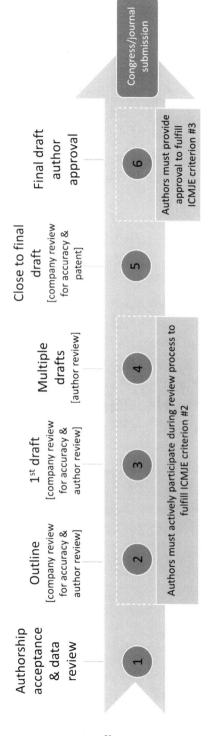

Following author acceptance of participating in publication development and data review, an outline is prepared that includes key structure and points to be included in the manuscript along with suggested figures, tables, and references. This may be prepared by an author or by professional medical writer under the direction of lead author(s). Company review is performed to ensure accuracy of information included in the draft, while authors review to provide input on the content. Following this, a first draft of the full manuscript, including introduction, methods, results, discussion, and conclusions is prepared. Some authors prefer to write the introduction and/or discussion and conclusion sections on their own, while the medical writer may prepare a draft with the methods and results sections. In such cases, the lead author or the author who agreed to write the introduction, discussion, and conclusion sections will complete the full manuscript draft prior to review by the company and all authors. This follows similar company and author review as done for the outline review. The manuscript can undergo multiple draft revisions with multiple rounds of review by the authors. Often, during these reviews, authors may request additional analyses or datasets be included in the manuscript. A close-to-final draft undergoes company review for accuracy and patent attorney review to ensure filing of patent (as needed), followed by final draft approval by authors prior to submission to the scientific congress or journal.

Author participation in providing a critical review of the draft and then final approval prior to submission are critical to satisfying the second and third ICMJE authorship criteria. These activities require documentation, and if authors fail to fulfill either one of these criteria, their name may be removed from author byline, which is in line with preventing guest authorship. According to the Council of Science Editors, guest authorship is defined as:

> *Authorship based solely on an expectation that inclusion of a particular name will improve the chances that the study will be published or increase the perceived status of the publication. The "guest" author makes no discernible contributions to the study [or publication], so this person meets none of the criteria for authorship.*

The name of an author is considered for removal generally after due diligence to obtain author feedback and approval and following consultation and agreement from the lead author (and/or SSC or PSC) to remove the noncompliant author's name. It is important for participating authors to be aware of and understand the consequences for noncompliance to the requests for draft review and author approval. Another common practice among authors is to simply reply with "Draft looks good" or "I am happy with the draft" without providing any meaningful comments or feedback during draft review. This raises concerns and can be argued to not satisfy ICMJE criterion #2. This is an area where lead or senior authors can take an active role in reaching out to the complacent participating authors and reminding them to provide more critical and meaningful review and feedback. In my experience, whenever lead authors have intervened, the complacent participating author almost always responds with a more thorough review and feedback on the draft. Here again, it's all about the mindset and behavior change. Authors must take the responsibility and ensure commitment to fully participate during the publication review and approval process.

A counter-argument to this has been proposed that because the lead and senior authors may have put in significant effort in preparation of the initial draft of the publication prior to circulating to the remaining authors for their critical review and feedback, the draft may be so well prepared that there is no further need for improvement, hence the remaining coauthors are likely to respond with "no comments" or "looks good." Although the draft may be in good shape, in my experience, as mentioned above, I have almost always seen coauthors come back with important insights and feedback after the lead or senior authors have pushed for the authors to provide critical review and feedback.

After approvals are obtained from all authors, a submission package is prepared that includes a cover letter, final draft of the manuscript, any supplementary materials required or planned to be submitted, and tables and figures according to the journal's specified format. Submission is completed either via email or online (more common), depending on journal requirements.

Overall, this entire process from the time when authors are engaged to time of submission can generally be as little as two to three months (occurs very rarely if all authors and reviewers are fully engaged and responsive) to as much as six to twelve months or longer (more common scenario). It has been reported that, on average, authors may spend 90 to 100 hours to prepare a scientific manuscript for journal submission (Ware and Mabe, *The STM Report*, 2009). For congress abstracts, the draft review and approval process is much shorter, from two to four weeks, while that for poster and oral presentation can range from four to six weeks. These are estimated time frames, typically, for industry-sponsored research publications, which have the added complexity of coordination and alignment between a company study team and academic study investigators. For non-industry-sponsored research, it may take less time for publication development.

Furthermore, the time it takes to prepare a manuscript is dependent on many factors: number of authors involved in a given publication, availability of all necessary data for the manuscript, response from authors during draft review and approval, and complexity of data (which may require multiple discussions to gain alignment).

Now we have completed a manuscript draft and submitted to a journal. But wait, we are not done yet! After submission to a journal, now the ball is in the court of the journal with more processes that are covered in the next chapter.

The Journey of Journal Publication: from Submission to Publication

If a publisher declines your manuscript, remember it is merely
the decision of one fallible human being, and try another.
– Sir Stanley Unwin

Once a manuscript is submitted to a journal, it's fate is decided by the journal editors and peer reviewers. As mentioned in Chapter 1, Kendall Powell (*Nature*, 2016) beautifully describes the plight of a young researcher, Danielle Fraser, who tries to get her manuscript published and painfully discovers the long and arduous process of submission and rejection multiple times before it finally gets published 23 months after the paper was first submitted to a journal. Fraser confirms that although the long peer-review process and revisions improved the paper, it didn't change the main conclusion of the paper. Indeed, one of the reasons that academia remains strongly committed to and researchers support peer review is that peer review is believed to improve the quality of a published paper (Ware and Mabe, *The STM Report*, 2009). In this chapter, we will walk through the steps of a manuscript's journey once its submitted to a journal.

Journal Editorial Review

Once a manuscript is submitted to a peer-reviewed journal, it will undergo screening for appropriateness of the topic or research as related to the journal's scope, style, completeness of information, plagiarism, etc. by the editorial office and/or journal editor prior to peer review. The manuscript can be rejected immediately by the editorial office even without going through the peer-review process or, if deemed satisfactory for peer review, it will be circulated for that process. Authors can receive rejection notice within one to two weeks after submission if the manuscript is rejected outright without peer review.

External Peer Review

If a manuscript is considered satisfactory, then the journal will circulate the draft to peer reviewers who are scientific experts in the area relevant to the publication. Peer review invitations are often sent to many reviewers, while a few may agree to complete the review based on availability and interest. In general, each paper is aimed to receive completed peer-review reports from two to three (in some cases more) reviewers.

Journal Editorial Decision Following Peer Review

Following peer review, in consultation with the peer reviewer feedback and recommendations, the journal editor decides whether the manuscript is rejected or requested to be resubmitted with revisions based on peer-review comments. Manuscript acceptance without revisions is extremely rare. The journal response time for a decision varies considerably across journals; in general, it can take about two to three months or longer. Depending on the extent of peer-review comments and required revisions, it may take a few weeks to a month or longer for the authors to respond, as all authors will need to review and provide final approval on the revised version prior to resubmission. Longer time is needed for revisions if peer reviewers requested additional analyses to be included in the manuscript. The author response should include a point-by-point response to each reviewer's comments, along with a revised version of the manuscript with marked changes. The revised manuscript may undergo another round of journal peer review prior to final acceptance and additional revisions may be requested based on peer-review comments. Each additional request for revisions leads to further delays in final acceptance. Once accepted, the paper may be electronically published online within a week or a few weeks from final acceptance for publication, followed by print publication typically months later.

Overall, half of the manuscripts submitted to a journal typically get rejected. Before submitting to a new journal, the paper is usually required to be reformatted to the new journal's specifications. This reformatting is done every time the manuscript gets rejected and then submitted to another journal, in order to meet the new journal's specifications and requirements. This is an area where we could have more efficiency in saving time, effort, and money, as the journal-specific formatting requirements are generally related to cosmetic aspects and length or word count of a manuscript. If journal publishers and editors aligned and agreed on a standardized manuscript format and length that is acceptable to all or most journals, it would alleviate the need for reformatting with every new journal submission. Alternatively, it would be helpful if the publishers agreed to receive the manuscript in whatever format it was previously prepared, then require the authors to format per the journal's specifications after it is accepted for publication.

Collectively, from the time authors are engaged to the time the paper gets published, it can take minimum of nine to twelve months up to two years or longer. A number of studies, as discussed in Chapter 3, have reported a median of about 24 months to publish! The process described here is for a single paper! Imagine this process multiplied by hundreds and even thousands of research articles (for large pharmaceutical companies) across multiple drugs and multiple studies. Based on my experience in pharmaceutical industry, collectively, companies spend millions of dollars per year to support preparation of manuscript drafts for initial journal submission with additional millions of dollars required to address the reformatting after each rejection and for the multiple rounds of revisions requested by the journal. This is where resources and budget constraints come into play. And as a community, we wonder why it takes so long to publish and why so many studies remain unpublished! The amount of time, energy, resources, and cost to publish in a peer-reviewed journal is enormous.

There are so many players involved in the publication process. If each player reflected on their contribution and tried to be more collaborative— sponsors and authors committed to preparing and completing publication drafts in a timely fashion, journal publishers improved efficiencies by removing unnecessary requirements (such as manuscript formatting for example) and streamlining journal review process—then it might bring us closer to achieving the goal of publishing all clinical trials.

Interactions Between Authors and Scientific Journals

To improve editorial standards, it is essential to understand the current status quo and obstacles facing journal editors and others in the peer-review and publishing process.
– Lorraine Ferris

Journal editors play a critical role in the selection of which submitted manuscripts will be accepted for publication in their journals. Development of a journal manuscript itself can take many months; however, its fate is only determined after the manuscript is submitted to the journal as described in the previous chapter. As noted in the above quote by Dr. Lorraine Ferris, past president of the World Association of Medical Editors (WAME), to appreciate the layers of complexity and impact, it's crucial to understand the current "status quo and obstacles" that are not only faced by journal editors but also by researchers and publication professionals when interacting with journal editors during the publishing process. There are numerous touch points when authors interact with journal editors or the editorial office: presubmission inquiry, submission, peer-review response, resubmission, acceptance, galley proofs, publication fees (if applicable), and post-publication (for digital media, if offered by journal).

Bhakti Kshatriya, PharmD

Presubmission Inquiry

Although not a routine practice for all manuscripts, authors often submit a formal presubmission inquiry to the journal up front to gauge the journal's interest in the research and to help make an informed decision on their journal selection. This also helps the authors avoid losing time and effort in the submission process in case the journal is not interested. Typically, a short summary of the study design and results along with a brief history of prior congress presentation (if applicable) or prior journal submission is provided to the journal within the inquiry. The journal's editor or editorial office reviews the inquiry and provides a response in writing as to whether the study is suitable for the journal's readership, if the research is of interest, or if the manuscript could be considered for publication in their journal. In some instances, the journal may inform the authors that the study or research is not suitable for the journal and that the authors should consider other journal options. The latter feedback during presubmission inquiry, however, is infrequent. Authors are often encouraged to submit their research to the journal only to later find their manuscript rejected by the journal even without undergoing the peer-review process.

Over the years, the practice of presubmission inquiry has been primarily based on anecdotal experience and is generally considered a helpful and potentially timesaving strategy. Boorer et al. recently presented their data based on one company's experience at the eleventh annual meeting of ISMPP in 2015: out of 48 presubmission inquiries submitted, 71% received a positive response from journals requesting the authors to proceed with full submission of the manuscript. Of the manuscripts that were submitted following a positive response, one-third ended up being rejected by the journal. When papers are rejected, it not only leads to disappointment and distress for the authors in having the manuscript reformatted for another journal, but it also contributes to a loss of time, energy, and money expended for the first submission.

One might question how there is a financial impact. While preparing the manuscript, either the author or a professional medical writer/editor are required to devote time to formatting the manuscript per journal specifications. Costs can be incurred based on clinicians' salary or fees related to medical writing assistance. Although positive response to presubmission inquiry cannot be expected to guarantee final acceptance for publication, it would be helpful to have more thorough and realistic journal feedback at the presubmission stage to avoid unnecessary effort and financial consumption related to submission and resubmission following a rejection.

Let's review some potential factors that could be contributing to journal rejections following a positive presubmission response. In order to maintain a certain profile of the journal with regard to rejection rate, it is of benefit to the journal to have as many manuscript submissions as possible even though they are likely to be rejected eventually. To counter this, an argument from the journal editors could be that the information provided in the presubmission inquiry may be insufficient to fully assess the suitability and hence they end up rejecting the manuscript later when the full paper is available for review. This can be questioned especially in instances when the reason for rejection provided by the journal is "unsuitable for our readership" or "data not novel." Journals should be able to make a definitive assessment on suitability for readership from a presubmission inquiry as long as there is sufficient information provided, including a summary of study design and brief results. Journal editors can also independently research online about the clinical study on public trial registries during the presubmission inquiry review.

Submission of Manuscript

Manuscript writing follows the established IMRaD format: **I**ntroduction, **M**ethods, **R**esults, and **D**iscussion are mandatory sections for all manuscripts. In addition to these four key sections, manuscripts also include an abstract at the beginning, summarizing the entire IMRaD in approximately 500 words or so. Although IMRaD is the general structure that is widely accepted for scientific communication, journal publishers have developed multiple variations. For example, some journals may require the methods section at the end of the paper, while some require the results and discussion be combined into one section. Still others have a separate **C**onclusion section at the end of the paper. This creates a wide range of possibilities, such as IRDaM, IMRaDC, IMRMRMRD, ILMRaD (L stands for literature review), and so forth. As previously mentioned, there are over 28,000 scientific journals, and each journal typically has individualized specifications regarding the format of the manuscript text, allowable tables, figures, word count, referencing, etc. This variation exists even with journals that are under the same publisher. Journal specific format allows each journal to have its own unique identity and presentation style in the final printed or published piece. Some journals that are owned by a parent publisher offer the option to authors of transferring the manuscript to sister journals along with peer review (which is also referred as cascade review). For example, if journal A rejects the manuscript, the editor may offer the authors the chance to have their manuscript automatically transferred (without the need for reformatting) to their sister journal B that is owned by the same publisher. However, if the sister journal B is not of interest to the author and the author wishes to submit to a different journal, he/she will need to reformat the manuscript per the new journal's specification prior to submission. This practice of journal-specific formatting has been driven by the journal publishers and editors, and the researchers have simply accepted it as "just how it's done."

Manuscript writing also involves standardization of data reporting according to study design, methodology, and type of data. For example, the CONSORT guidelines, which are most widely recognized and endorsed by many medical journals, are specifically for reporting results of randomized controlled trials. Such guidelines are extremely helpful in ensuring consistent reporting of data. Below is a list of suggested guidelines for various types of data reporting:

- **CHEERS:** Consolidated Health Economic Evaluation Reporting Standards: http://www.valueinhealthjournal.com/article/S1098-3015(13)00022-3/pdf
- **CONSORT:** Consolidated Standards of Reporting Trials: www.consort-statement.org
- **EQUATOR:** Enhancing the Quality and Transparency of Health Research: www.equator-network.org. Website provides multiple guidelines.
- **MPIP:** Medical Publishing Insights & Practices initiative: www.mpip-initiative.org. Website provides multiple guidelines.

- **PRISMA:** Preferred Reporting Items for Systematic Reviews and Meta-Analyses: www.prisma-statement.org
- **STROBE:** Strengthening the Reporting of Observational Studies in Epidemiology: www.strobe-statement.org

Peer-Review Process

Peer review is a vital component of scientific publishing as a form of quality control mechanism and tool to confer credibility. Within the medical community, scientific research will not be considered credible until it has been validated by peer review. The number of peer reviewers for each manuscript can vary, ranging from two to three or more reviewers, and can consist of medical experts within the area of research being reported, statisticians, and even patients. It serves several key objectives:

- Provides authors an opportunity to receive initial reaction and feedback from peers before the publication is widely available
- Allows authors to receive expert input on the research being reported
- Helps with selection of quality research (filters out poorly conducted studies)

During the peer-review process, journal editors often also take into consideration whether there is an overlap in the research with manuscripts that are already in press or published. This is important to ensure that there is no duplication of publication of same study results (i.e., duplicate publication). However, in certain instances the paper may get rejected simply because another research group has published a similar type of study, hence lack of novelty. Generally, publishers manage the peer-review process with internal infrastructure, while there are outsourcing options also available with independent peer-review firms that cater to authors and publishers.

With a boom in scientific research, there is an increasing demand for peer reviewers to process the growing volume of publications. Researchers generally consider it their duty and part of their job to peer review the work of others. Due to increasing demands, younger scientists with less experience than the more seasoned experts are being utilized as peer reviewers. The demand for peer reviewers will likely increase further with the introduction of the data sharing initiative when more secondary publications are expected to be published. Given less experience, it is important to ensure that less experienced peer reviewers are trained on how to provide critique for draft manuscripts and how to maintain quality of review. Although some journals provide required training for peer reviewers, this is not general practice within journal enterprises, and often, novice peer reviewers are left on their own to train themselves. Some journals provide a checklist or list of questions to address during a peer review. Potential peer reviewers can also refer to COPE's Ethical Guidelines for Peer Reviewers (which can be found on www.publicationethics.org) for general principles and guidance on the peer-review process and preparing peer-review report.

In addition, providing peer review is an altruistic act. In a survey, reviewers reported to spend on average nine hours (median five hours) to review a paper (Ware and Mabe, *The STM Report,* 2015). However, peer reviewers are routinely not compensated nor given credit or recognition in the publication for their time and effort in completing the review. A few journals are using innovative ways to compensate. For example, Veruscript (journal publisher) and *Collabra* (an open access journal) offer peer reviewers a portion of the journal's revenue with an option to keep the payment for themselves or contribute to a fund to help other authors who may not be able to afford publication fees. This is an interesting concept of giving back to the research community. Moreover, peer reviewers are not as interested in monetary compensation but would prefer to receive some level of recognition and credit for their effort.

Traditional journals generally follow a *closed review process*, meaning the comments provided by the peer reviewers are only shared with the authors and never publicly disclosed. Furthermore, the process can be single-blinded (reviewers know the names of the authors, but the authors do not know the reviewers' names), double-blinded (neither the authors nor the reviewers are aware of each other's identity), or open (identity of authors and reviewers are known to both parties). There are advantages and challenges to each of these approaches (see table). Open peer review offers increased accountability, fairness, and transparency. With the increasing call for transparency, newer online publishing platforms such as *F1000Research* follow an *open, non-blinded peer-review process*, whereby the initial manuscript is published online within days of submission, followed by open peer-review process allowing full visibility of assigned reviewers and their feedback, as well as subsequent revisions submitted by the authors. Some journals, such as the *BMJ*, certain *BioMed Central*-series journals, and others, have also adopted open peer review and publish the prepublication peer-review history along with the online published article.

Type of Peer-Review Process	Advantages	Challenges
Single-blinded • Reviewers are aware of author names; authors are not aware of who the peer-reviewers are • Most common form of peer review among life science journals	• Reviewers are able to provide feedback more candidly without concerns for criticism from an author	• Knowledge of the authors can result in either preferential review (if reviewer is amiable towards the author) or in undue excessive criticism or discrimination (if reviewer dislikes the author) • Lack of knowledge of the reviewers does not allow authors to appeal editorial decision (if there is known animosity from the reviewer towards the author which the journal editor may not be aware of; or if the reviewer may not have sufficient expertise)

Type of Peer-Review Process	Advantages	Challenges
Double-blinded • Neither authors nor reviewers are aware of each other's identity • Common form of peer review among social science and humanities journals	• Potentially unbiased review • Both authors and reviewers are protected against criticism from each other	• Anonymity is not fully guaranteed – reviewers may be able to identify authors based on area of research and prior knowledge of study involvement, resulting in similar concerns as above • Similar concerns as above for authors due to lack of knowledge of the reviewers
Open • Identity of authors and reviewers are known by both parties • Reviewer feedback may or may not be published together with the article • Some journals have adopted this model	• Full transparency of review allows accountability and more civilized review with appropriate tone of feedback • Generally, improves overall quality of review or there is no negative impact on quality of review • Can offer a collaborative environment for reviewers and authors to improve the research work / publication • Reviewers are more motivated to do a thorough job and suggest more justified/ fair changes or additional data requests, especially if the reviews will be published together with the article	• Some reviewers may decline to participate based on open system, due to concerns of scrutiny in case negative feedback is provided on the research • Reluctance to criticize work of more senior scientists, especially if there is potential for direct impact on reviewer's career

There are three additional variations of peer review, including cascade review, portable review, and post-publication review. Cascade review avoids the need to repeat peer review each time the paper is rejected. Upon author's permission, the manuscript and accompanying review reports are transferred to the participating new journal. This approach has been used by *BioMed Central* journals and *PLoS ONE*. Portable peer review is similar to cascade review, where the authors have the option of taking the reviewer report and submitting to a new journal that accepts it. This also applies to situations when peer review may be completed by an independent peer-review service that caters to authors and publishers. Rubriq is one such peer review service, which charges a fee to the authors or publishers, while Peerage of Science is a free peer review service for authors (its revenues come from publishers, funding organizations and universities). Post-publication review can include selected reviewers or general readers as reviewers. For example, this type of review can be found on PubMed Commons and individual journal websites such as the *BMJ*, the *NEJM*, journals by Frontiers Media S.A, and others.

There has been an increasing criticism for potential bias in peer review in recent years (Lee et al., *JASIST*, 2013; Rennie, *Nature*, 2016). Moreover, the most common form of peer-review process currently used is closed, single-blinded process among medical science journals. This is an area where journal publishers are putting in a lot of effort in exploring innovative ways to address the growing criticisms of the traditional peer-review process, including ineffective reviews; bias, especially with the single-blind process; and inefficiencies leading to delay in publication. Some journals are experimenting with a double-blind process, which doesn't appear to fully address the issue of bias since the reviewers are able to identify authors on their own based on prior knowledge of research study, prior publication history, reference list, etc. Although traditional journal publishers remain skeptical about open peer review, there seems to be growing interest in this model. Concerns related to open review appear to be partly due to not knowing what the impact of making such a change might be: how well or poorly would such a change be received by reviewers? Will it impact the quality of review provided by peer reviewers? Will it result in spurious criticism that is motivated by commercial interest, competitiveness, or academic jealousy? These are some reasonable concerns, and a number of studies have reported lack of support or preference for open peer review, and peer reviewers' unwillingness to participate in open review process (Lee et al., *JASIST*, 2013). However, several other studies have also reported usefulness and no negative impact on quality of review with open peer review (Freire, *Political Methodologist*, 2015; Lee et al., *JASIST*, 2013). Here I have highlighted a few. Hopewell et al. (*BMJ*, 2014) investigated the impact of open peer review in 93 articles reporting randomized controlled trials in open review *BioMed Central* (BMC)-series journals. The study found that the number of changes requested by peer reviewers was relatively small and most were considered to have a positive impact on the final manuscript, while only 15 out of 93 manuscripts were required to provide additional unplanned analyses by the peer reviewers. Furthermore, randomized studies conducted by the *BMJ* (van Rooyen et al., *BMJ*, 1999; van Rooyen et al., *BMJ*, 2010; Groves, *BMJ*, 2014) have found that removing anonymity of reviewer and author names "improved the tone and constructiveness of reviews without detriment to scientific and editorial value," and informing peer reviewers about public access to their reviews did not negatively impact the quality of the reviews. Another randomized controlled study reported an improvement in the quality of peer review with open process (Walsh et al., *British Journal of Psychiatry*, 2000).

Another criticism of peer review has been the continued practice of rejecting papers because of lack of novelty or interest. This is based on a very subjective opinion of the journal editor and/or peer reviewers and can undermine the true importance and scientific value of the research. As mentioned previously, Albert Einstein's papers underwent multiple rejections before they were accepted, which included seminal findings that later led to his theory of relativity and discovery of the law of photoelectric effect that won him the Nobel Prize in physics (Isaacson, *Einstein: His Life and Universe*, 2007). Economist George A. Akerlof's work and paper on "The Market of Lemons" describing the concept of asymmetric information (decision making based on one party having more information) was rejected three times because of being trivial before it was accepted and published. He later received the Nobel Prize in economics for this and his later work (Akerlof, 2003).

In the 2006 study published in the *Medical Journal of Australia*, Lee et al. reported that 70% of 1,107 manuscripts submitted to three top-tier journals—the *Annals of Internal Medicine*, the *BMJ*, and the *Lancet*—were rejected outright without undergoing peer review and 24% were rejected after peer review. This study did not investigate or report the reasons for rejection provided by the journal, although lack of novelty or interest can be speculated for outright rejections. The authors of this study later investigated the journal rejection/acceptance and eventual citation rates for manuscripts submitted to the same three journals (Siler et al., *PNAS*, 2015). Of the 1,008 manuscripts submitted, 946 (94%) were rejected, of which, 757 (80%) were eventually accepted and published by other journals. Of the rejected manuscripts, 772 (82% of the rejections, and 77% of total submissions) were rejected outright by the initial editorial review without undergoing peer review. Further analysis of the citation rates for rejected manuscripts showed that although manuscripts with outright rejection tended to have lower average citation rates than those that were sent for peer review, 12 of the 15 most cited papers had been rejected outright by the initial journal. The study also reported that lack of novelty was one of the common reasons for rejection. Interestingly, among the most highly cited papers, about 30% of the papers were published in journals with higher impact factor than the initial journal, while the remaining papers were eventually published in journals with lower impact factor compared to the initial journal. One of the limitations of this study is that it only evaluated papers that were initially submitted to three top-tier journals that are known to have very high rejection rates. It is likely that the initial or overall rejection rates with mid- or low-tier journals would be lower than those observed in this study.

In summary, most life sciences journals follow closed, single-blinded peer-review process, while some such as the *BMJ* and certain *BioMed Central* journals have been using open peer review, and others such as *PLoS ONE* give their reviewers an option to be identified or remain anonymous. With more experience, it is hoped that journals will gain more confidence in the open peer-review model to make the transition. We are now in the era of transparency, and open communication will be the hallmark of scientific publishing for future generations. The practice of accepting journal articles based on subjective assessment of novelty needs to addressed, while a few open access journals such as *PLoS ONE* have recognized this issue and adopted their editorial and peer review based on "soundness of science" rather than novelty of research. Peer reviewers have the option of archiving, sharing, and tracking their review reports on an online platform—Publons (www.publons.com). It also allows members to link to their ORCID profile, thus allowing researchers to keep record of their research and publication activities.

Galley Proof Review

Authors are required to complete a galley proof review, often with a very short turnaround time, prior to the publication of their article in print.

Post-Publication

Health care professionals (HCPs) have diverse preferences for learning and receiving information. With technological advances, many journal publishers include a variety of digital media to accompany and complement the published article, such as video abstracts, slide presentations, mechanism of action videos, and author interviews. These tools offer a broad range of scientific communication for reaching a wider audience and meeting the educational needs and preferences of health care professionals.

Social media, via platforms such as Facebook, Twitter, and LinkedIn, have allowed broader engagement of health care professionals to express their views and opinions and to discuss published literature. A number of platforms such as PubMed Commons and certain journal websites also provide readers the opportunity to comment on published articles. PubMed Commons offers a centralized platform across journals for reader feedback, however, only existing PubMed authors or those invited by PubMed authors can provide feedback on published articles.

Sharing Patient-Level Data in the Era of Transparency

What data sharing is really about: getting more and better answers from the data [that] patients have provided in clinical trials in order to benefit other patients.
– Charlotte J. Haug

Sharing patient-level data has been a topic of great interest and debate among regulators and the industry for over a decade. The debate has more recently expanded to include the wider research community following the announcement by ICMJE of their data sharing proposal, in January 2016. Why should we care about access to patient-level data? Benefits of data sharing can be categorized into two main buckets:

- **Ethical Obligation to Research Participants.** Sharing of research data for reanalysis can reduce the potential of duplicative studies, which could subject participants to unnecessary exposure to previously tested interventions and hypotheses. Furthermore, as noted in the above quote by Dr. Charlotte J. Haug (international correspondent for the *NEJM*), patients participate in clinical trials in part to provide data to benefit other patients, and if the data is not widely shared, then that benefit is not likely to occur or will be missed.

- **Enhancing Scientific Integrity and Medical Progress.** Reconfirmation through reanalysis of previously reported data and performance of a risk/benefit analysis allows the scientific and medical community to build trust in the reported research and confidence in the tested intervention, which can ultimately result in better patient care. More importantly, use and application of previously collected data to answer new scientific questions and theories carries a huge societal benefit of financial savings, avoidance of undue treatment exposure to patients, and speed of knowledge by alleviating the need to conduct new studies to answer questions. Therefore, access to data can accelerate research, enhance collaboration, and help build trust within the scientific research community.

In 2003, the NIH issued its policy for data sharing, requiring investigators who seek $500,000 or more in direct costs in any single year to address data sharing plans in the grant proposals as a condition of receiving funding. In October 2014, the European Medicines Agency (EMA) issued its policy requiring biopharmaceutical manufacturers to make individual patient-level data available for medicinal products that receive marketing authorization. Prior to the EMA policy, GlaxoSmithKline (GSK) was the first pharmaceutical company to launch an online platform, in May 2013, for researchers to request access to anonymized patient-level data from GSK-sponsored trials. Building on their experience, other pharmaceutical companies joined the data sharing initiative, which has led to the launch of a single, multi-company platform (www.clinicalstudydatarequest.com), in January 2014, including 13 companies. Some companies used a different approach by collaborating with academic institutions—for example, Johnson & Johnson partnered with Yale University (The YODA project, www.yoda.yale.edu), and Bristol-Myers Squibb partnered with Duke Clinical Research Institute (Supporting Open Access to Researchers [SOAR] initiative)— to share their clinical trial data, while others have independently created their own data sharing platforms. EMA's data transparency policy helped biopharmaceutical companies prepare for data sharing programs that will no doubt have wider usage in the future.

Furthermore, private funders such as the Bill and Melinda Gates Foundations have also issued policies supporting data sharing, where researchers are required to make data accessible as a condition for receiving the research funds. More recently within the past year, this topic has sparked a wider discussion and debate among journal editors, the industry, research funders, and academic researchers following ICMJE's proposal requiring authors to share patient-level data underlying the results reported in the research article in order to be considered for publication in their member journals. Since the ICMJE announcement, there has been a flurry of discussions and debate in countless blogs, over a dozen editorials in the *NEJM* alone, and many more across other scientific journals. The reaction from the medical research community and industry in general has been unanimous in support of clinical trial data sharing in principle. However, academic researchers have expressed strong and often polarized opinions and disagreements mainly related to *how* data sharing should be done.

ICMJE announced its data sharing proposal in January 2016, and it was open for public feedback until April 2016. As a condition of consideration for publication in its member journals, ICMJE proposed the following requirements for authors:

- Share anonymized patient-level data underlying the results reported in the article within six months after publication
- Effective for clinical trials that begin patient enrollment one year after ICMJE adopts data sharing requirements
- Include data sharing plan as a component of clinical trial registration. Plan must include proposed location of where the data will be housed and mechanism by which data will be shared (if not in public repository)
- Include description of data sharing plan in submitted manuscript
- If there is non-compliance to data sharing after the article has been published, then journals may request additional information from the authors; "publish an expression of concern;" notify the sponsors, funders, or institutions; or retract the publication, if necessary
- Exception may be granted in rare cases and reasons for the exception must be provided in the publication

ICMJE received an overwhelming response to the proposal with 320 commentaries from researchers and organizations across the globe. The feedback focused on the overall requirement to share data, requirement for data sharing plan, six-month time frame for making the data available, and the proposed approach for providing credit to the original researchers, all of which are publicly available on the ICMJE website. Subsequent to this, the *NEJM* published several editorials reporting select proposals and feedback that were submitted to the journal by researchers and organizations. Some highlights of the feedback and/or proposals are described here.

- **Creation of a global, centralized data sharing portal (Vivli)** connecting industry, academia, government and non-government organizations, and the existing systems (Bierer et al., *NEJM*, 2016). Development of this portal is sponsored by the non-profit organization Multi-Regional Clinical Trials Center of Brigham and Women's Hospital and Harvard (MRCT Center). The portal would aid in implementing the data sharing initiative through secured data hosting; provide data request services, including an independent review panel; and include granular search functionality based on curated data from existing repositories. The portal will link existing data sharing platforms and communities, as well as host data from investigators who do not have the resources to share data. Hosted patient-level data will be required to be anonymized, and researchers who generate the data may be able to receive assistance from Vivli directly for anonymization or receive contacts for anonymization service providers. Vivli will monitor if data anonymization is sufficient through statistical sampling. Data requesters will be required to execute a data use agreement (DUA) in order to receive access to data and assure commitment to publish the secondary analysis results, agree to not share data beyond what is specified in the agreement, and agree to not reidentify the study participants.

Datasets will be tracked with unique identifiers (e.g., digital object identifiers [DOI]), and linked with the data generators via ORCID (Open Researcher and Contributor ID) identifiers and with publications that cite the data. The exact launch date of this platform has not been announced at the time of writing this book; however, the organizers plan on partnering with existing data sharing systems to expedite platform development.

- **Use of existing data sharing platforms** established by pharmaceutical companies such as www.clinicalstudydatarequest.com and other individual company data sharing platforms; collaborative systems such as the Yoda Project and SOAR initiative; and academic consortia such as the Academic Research Organization Consortium for Continuing Evaluation of Scientific Studies – Cardiovascular (ACCESS CV). Leveraging the use of existing data sharing platforms is a practical approach. However, it leaves the information scattered among individual sites. Hence the above proposal for Vivli as a single portal linking various platforms is an ideal and optimal solution to provide researchers with a convenient one-stop-shop for data sharing.

- **Encourage a collaborative, symbiotic approach for data sharing.** An example of this has been reported for the Worldwide Anti-Malarial Resistance Network (www.wwarn.org) that includes 260 collaborators from 70 countries, with the data repository containing clinical trial data generated by academic institutions and pharmaceutical companies (Merson et al., *NEJM*, 2016). WWARN data generators are allowed to be fully involved in the process of any meta-analysis or secondary analyses and are recognized on publications in accordance with ICMJE authorship guidelines. Another reported example is the experience of the Bill and Melinda Gates Foundation for the Healthy Birth, Growth, and Development – Knowledge Integration (HBGDki) initiative, which has generated data repository from 420 clinical and population survey studies from 50 countries (Jumbe et al., *NEJM*, 2016).

- **Use of data enclaves with distributed analysis methods** that allow investigators to conduct analyses without taking control of the data. This approach has been proposed particularly in response to challenges associated with data anonymization for certain study methods, such as cluster–randomized trials that are routinely done for health care data (Platt and Ramsberg, *NEJM,* 2016). In such studies, there are no individualized research subjects; instead, data are randomized according to health systems or providers. The data can involve the diagnostic, procedural, or treatment intervention information of thousands of patients who do not give explicit consent. It is often impossible to de-identify providers or organizations and has an increased risk for unauthorized reidentification. Data enclaves with distributed analysis methods have been used successfully by the Centers of Medicare and Medicaid Services Virtual Research Data Center.

- **Academic reform for researcher credit, tenure assessments, and funding reviews**. There is growing approval among the research community as well as by ICMJE for establishing an alternative method of providing credit to researchers within the academic community. Researchers who originate the data (also referred as data generators) should receive appropriate credit whenever secondary analyses are conducted and reported by researchers who have obtained data through data sharing (also referred as data recipients). The current model of authorship on published papers may not be appropriate or possible for giving credit to data generators, especially when the secondary analyses are done independently without collaboration with data generators. ICMJE has welcomed suggestions to address this. Lo and DeMets (*NEJM,* 2016) suggested that data generators could receive academic rewards when other researchers use their shared data to publish secondary papers, and research centers and universities may reward data sharing in hiring and promotion decisions. They also suggested to have shared authorship in cases of collaboration; however, this can be challenging, especially when data is being shared from large multicenter trials or when data is pooled from multiple studies involving hundreds of investigators. In such cases, only a handful of original investigators (data generators) would be able to participate and qualify for authorship. In order for remaining investigators to get credit for data sharing, it would be appropriate to create a new standardized category of credit on publications, specifically labeled as data generators. The list of investigators can be provided as supplementary material, while ensuring that their names are linked to the publication.

- **Ensure a peer-review process to assess scientific validity of data request proposals.** In general, there appears to be consensus among various stakeholders on this. This has already been implemented in the established data sharing programs set up by biopharmaceutical companies for industry-sponsored clinical trials, where an independent review panel (IRP) reviews and approves the proposal prior to granting access to data. The Vivli project organizers and ACCESS CV members also suggested similar IRP requirements.

- **Execution of data use agreements (DUA) as a condition for data access.** Various stakeholders also seem to be in agreement with this. A DUA would include a commitment by data requesters to publish the results of secondary analyses, and ensure appropriate handling and use of data as described in the agreement (e.g., no reidentification, no sharing of data beyond that specified in the agreement, etc.). Again, the data sharing platforms established by biopharmaceutical companies require data requesters to execute a DUA after IRP approval and prior to granting access to data in a secure online environment. The Vivli project organizers and ACCESS CV members also proposed similar requirements for a DUA.

- **Secure web portal for data access and analyses.** Similar to the above points, here again, there seems to be overall consensus among various stakeholders to ensure that access to data is restricted to a secure setting. In addition, all anonymized data must ensure protection of patient privacy.

- **Extension of data access request to two to five years post primary publication** instead of ICMJE's proposed six-month time frame. This was proposed by the International Consortium of Investigators for Fairness in Trial Data Sharing (*NEJM*, 2016), along with 282 investigators from 33 countries, and received a huge outcry on social media from the research community. The basis of the argument by the Investigator Consortium was that the original researchers or data generators spend significant time, resources, and effort to collect the data and reserve the right to have adequate time to complete and publish primary as well as secondary analyses before opening the data for sharing with other researchers. The thought process behind this argument reminded me of another analogy—the introduction of generic drugs by generic manufacturers following patent expiration of drugs where the original manufacturers have spent millions of dollars, effort, and resources in developing the innovative medicine or treatment. After patent expiry, the generic manufacturers sell the drug and reap profits without any investment in its original research and development. Like the innovative drug manufacturers, researchers are now faced with a similar situation to give up data they have invested significant time, effort, and resources to collect. Ultimately, as a principle, the data are owned by patients, so in the case of data sharing, the research community owes it to the patients and hence to the public who volunteered to help generate the data and without whom there would be no clinical data at all. However, it's interesting to see the thought process for delaying access to data as proposed by the Investigator Consortium.

Moreover, it highlights the need for academic researchers to develop formal publication plans for clinical studies in order to complete and publish the primary as well as secondary analyses data of interest in a timely manner. Certain secondary analyses publications can be planned up front, even before the trial is completed, based on the study analysis plan. Formal publication planning is a common practice in the biopharmaceutical industry for industry-sponsored studies.

Furthermore, the six-month time frame proposed by ICMJE will definitely be inadequate for industry-sponsored study publications of investigational drugs that have not yet received regulatory approval or market authorization. For these publications, ICMJE needs to clarify and perhaps explicitly state in their final policy that such publications will not be within the scope of six-month data sharing requirements until the drug has received necessary regulatory approval or the drug has been terminated for development. This would also be consistent with the 2015 Institute of Medicine Report in which one of the goals for timing of data sharing is noted to "protect the commercial interests of sponsors in gaining regulatory approval for a product."

- **Researchers' concerns of abuse of study data.** The concerns of data abuse appear to generally stem from fear of releasing the data and not fully knowing what the data recipient might do with it. As an analogy, the Internet offers great benefit and convenience with online purchasing, where our personal information such as credit card details are widely exchanged, and there exists inherent risks and potential for hackers to misuse the data to their advantage. However, should that be precluding us from using the Internet for payment of services and goods altogether? What's important is to *acknowledge the potential risks* and *ensure adequate security measures are in place to prevent abusive situations or behavior.* Data requestors who are found to have abused study data or who breach a DUA can be tracked in a centralized database and permanently barred from receiving datasets in the future (similar to physician and investigator debarment lists that are publicly maintained by national regulators). Qualification of researchers can be addressed through unique identifiers such as ORCID as part of the data request application.

The highest risk of potential abuse is with freely available data on public repositories, where there is no mechanism to check the credibility of proposed scientific research, qualification of researchers, or their prior history. Therefore, it would be prudent to adopt the process of a formal data request application along with prior approval by a review panel and execution of a DUA as standard practice, while making data sharing on public repository as a secondary approach reserved only for rare, appropriate circumstances.

The biopharmaceutical industry has set precedence through their commitment and establishment of data sharing platforms where access to data can be requested for eligible studies following manuscript acceptance for publication. Initial data on the volume of requests and how shared clinical trial data are being utilized is available for the existing three data sharing platforms—clinicalstudydatarequest.com, the YODA project, and the SOAR initiative (Navar et al., *JAMA*, 2016). Requests were received from 17 countries, although more than half were from the United States. Of the total of 3,255 clinical trials available across the three platforms, only 15% have been requested for data since the inception of the platforms (earliest in 2013) up through December 2015. Of the 505 requested trials, 70% of the requests were for Phase III trials, which accounted for about 25% of total available Phase III trials. Validation of primary study results was noted for only 4% of requests, while the majority of proposals were focused on non-prespecified subgroup analyses or analyses for predictors of response. Of the 234 proposals submitted, 154 had received approval by review panel, 113 had completed data sharing agreements, and only one had been published at the time of this analysis (which was a validation assessment reporting contradictory results from the original research article). So far, it appears that the early use of data sharing platforms has been limited, which could possibly be related to lack of awareness of these resources or lack of funding to conduct secondary analyses. The utility of clinical trial data is likely to grow with increased awareness and funding for such analyses in the future and researchers could also gain more confidence in data sharing with more experience.

- **Lack of infrastructure, resources, and a standardized approach for the global research community.** One of the concerns raised is potentially widening the research–output gap between countries with low resources and no infrastructure and those with high resources. By making data sharing mandatory, researchers from high-resource countries will have the ability to use and reanalyze data collected by researchers from low-resource countries, but the reverse will not be likely due to lack of resources. To alleviate these concerns, a collaborative approach that involves the original data generators in the secondary analyses would be optimal. An example of such a collaboration has been reported for the anti-malarial WWARN initiative (www.WWARN.org), whose efforts have led to finding new trends and identification of subpopulation effects with greater certainty, which contributed to changes in global treatment guidelines for malaria (Merson et al., *NEJM*, 2016).

In summary, sharing of clinical trial data is imminent and essential to increasing trust in and transparency of clinical research, to fulfill our moral obligation to study participants, while leveraging existing data to help advance scientific discovery through stimulation of new ideas for research and avoiding unnecessary duplicative studies. Collaboration between all stakeholders—journal editors, funders, the industry, academic institutions, and researchers—is vital to its successful implementation. Policy makers need to carefully *listen* to the concerns that are raised, identify solutions through mutual and collaborative discussions, and help create a culture of responsible data sharing through adoption of best practices that are pragmatic and respectful of each stakeholder's needs and concerns. While it may not always be possible to have a collaborative working relationship between data generators and data requestors, a collaborative, symbiotic approach would be ideal and should be considered for making it a standardized practice among researchers. This would allow data generators to be involved in the secondary analysis, avoid data misinterpretation, and provide valuable insights to data requestors that might otherwise be difficult to obtain simply through receipt of anonymized datasets. This could then result in more enriched new research findings. The optimal approach for data sharing would be through a formal data request application by qualified researchers, approval by a review panel, execution of a DUA, and access to data in a secure setting to ensure scientific integrity and alleviate inappropriate use of the data that is being shared. Publication professionals and authors will need to consider data sharing requirements within the publication plans to ensure appropriate coordination and integration of data sharing activities.

The Path Forward

Reform can be accomplished only when attitudes are changed.
– Lillian Wald

As I reflect upon the transformation and recoding of our scientific publishing and data sharing culture, along with the technological explosion creating new communication norms over the last decade, I feel like we are at a crossroads and envision the creation of new possibilities and opportunities for further enhancements. These are exciting times to witness the fusion of medical publishing and technology.

Transformation of Social Communication Norms

Before we dive further into the path forward for scientific publishing, I would like to spend some time to have a look at what our current social communication norm is and how that has undergone transformation to dramatically affect not one but multiple generations within a short timespan: Generation X, Generation Y (or millennial), and Generation Z (or iGen). The reason for this comparison is that the way we communicate science should reflect to some degree the communication norms of that time. As I mentioned at the beginning of this book, the idea of periodical printed journals emerged out of need three centuries ago and was based on the communication norms at the time it was established.

With the introduction of the Internet, email, smartphone technology, apps, and platforms such as Facebook, Twitter, Instagram, etc., communication and exchange of information is now expected to be instantaneous, free, open, transparent, and with a broad reach. Collectively, this change has fueled and provided freedom of choice—the fundamental human right and basic need. It has completely revolutionized a wide range of industries such as music, telecommunications, and retail, which only two decades ago, not many people would have even imagined or believed could undergo such drastic transformation. Music, for example, was for decades owned and sold by entertainment companies by first selling records, then cassettes, followed by compact discs (CD). Consumers had to buy the entire record, cassette, or CD that would hold a set number of songs, even if the person only liked and wanted one or two songs. Steve Jobs's vision completely disrupted the music industry and how consumers purchase the music that they love through the introduction of the iPod. Today, even the CDs, which were the norm a little over a decade ago have become antiquated and nearly extinct. Now consumers can download and enjoy thousands of songs of their *choice* instantaneously, for a minimal price (around ninety-nine cents per song). Access to music is at the level of song, not the entire album!

Likewise, Amazon has completely disrupted the retail industry. Prior to Amazon, there were smaller incremental changes and innovations with regard to bulk purchasing that transferred into consumer savings. But the introduction of Amazon completely revolutionized the shopping experience altogether. It offers substantial incremental benefits of convenience, time and cost savings, and access to consumer product reviews while offering a massive selection of products that would be physically impossible for brick-and-mortar stores to carry in their inventory. Consumers are no longer required to physically go to a store, or worse, go store-to-store, to get a bargain or find the lowest possible price for the same product. Amazon offers the convenience of searching and comparing products, the ability to view and provide consumer ratings and feedback, and the ability to purchase the desired products instantaneously, along with having the option of fast delivery. Similar online shopping platforms have now become the norm not just for goods, but also for services, including hotel, airline, transportation, and many other industries. Manufacturers and service providers can no longer drive consumer decisions simply with marketing and promotional tactics. Consumer ratings and feedback will uncover and counter the marketing claims if they are unfounded; thereby improving transparency, quality, and honesty (i.e., manufacturers and services providers would need to be truthful about what they claim in their marketing materials).

Social media platforms such as Facebook, Twitter, Instagram, and others have alleviated our fear of providing mass communication without a filter and are grounded on complete transparency. They allow the users to not only widely share and communicate information openly, instantaneously, and at no cost, but they also allow users to express their opinions of others' work and information—by rating *like* and adding comments. Facebook created a whole new category of metrics—number of likes! Less than ten years ago, not many people would have fathomed even having such a platform or think they could give an opinion about someone else so freely and openly, yet this is now the norm!

As technology advances, consumer experiences and norms will continue to shift along with expectations. Current norms of communication have already established and conditioned generations of humans with the following minimal expectations:

- Instant exchange of information or communication
- Mass outreach
- Ability to express opinion on other's work or information
- Judge quality and make decisions based on peer rating and feedback (raising the bar for honesty and integrity)

Rationale for Recoding Scientific Publishing Model

Scientific publishing needs to ensure it can keep up with the quantum leaps made in technology and in science and medicine. Currently, the impact of entry of a younger tech-savvy generation into the academic and medical research field may not be as apparent since they are outnumbered by their seniors representing the prior generations. As expected, when starting a career, the younger generation will likely do whatever they are told as the established process or norm. However, over time, the workforce will continue to evolve, where tech-savvy individuals will become the majority and the basic needs and expectations of this newer generation will become more prominent. We need to ensure we have prepared and recoded our scientific publication systems to address those basic expectations. I would like to tie this back to the four critical elements that I described in the preface of this book: scientific integrity, transparency, access, and speed. All four critical elements, interestingly, fit well with the minimal expectations of how we currently communicate. *Scientific integrity* speaks to the importance of honesty and truthfulness; *transparency* is related to our ability to express opinions freely as well as honestly; *access* is linked to mass outreach; and *speed* is linked to the expectation of instant communication.

Scientific Integrity

Scientific integrity is the foundation of medical research and helps build trust among various stakeholders, including researchers, regulators, the industry, and ultimately patients and the wider public. Over the years, there have been many regulations, policies, and guidelines put in place to build and maintain scientific integrity and trust, including the Declaration of Helsinki, Good Clinical Practice, ICMJE guidelines, Good Publication Practice, COPE's code of conduct for journal publishers and journal editors, and others. Within scientific publishing, scientific integrity is, in part, protected and ensured through peer review, and hence it continues to remain as a cornerstone of the publication process. Yet we are at a time when there is continued public distrust in scientific research, which can be attributed to limited transparency, restricted access, and delay in communication and sharing study results and data. With the ongoing changes, we hope to see this improve in the future.

Transparency

Clinical Trial Disclosure and Reporting

As discussed previously, transparency in clinical research and publishing is essential and has received significant attention from the public, regulators, and research funders within the last decade, with increasing demand for transparency through clinical trial registration and results disclosure. The research community, in turn, has made significant progress in fulfilling these obligations that did not exist at the turn of this century. This progress, however, as discussed earlier, has been in response to legal obligations or policies of research funders. Most recently, finalization of this regulation in the United States along with the complementary NIH policy to further strengthen this requirement and improve compliance is a clear indication that further improvement is warranted in this area. It is hoped that the possibility of monetary penalties for non-compliance will help to further drive change in behavior and improve compliance.

Here I share some thoughts and strategies to further improve compliance with clinical research disclosure and publication, which I had previously revealed in one of my blogs on www.publicationpracticecounsel. com (posted November 2016):

1. **Changing Our Mindset.** Over the years, the medical research community has been faced with increased government regulations and more stringent policies in an effort to improve transparency and trust in medical research. However, it also raises the question— must there be laws and regulations to do what is simply an ethical obligation? If the entire medical research community, including the industry, academic institutions, and researchers, took accountability and responsibility to ensure transparency, then why would there be any need for laws and regulations at all?

    There are some who continue to believe *we will do what's required and do not need to commit to disclosing the types of studies that are not required.* Let's think about this for a moment. First, being in the health care industry, the main purpose for biopharmaceutical and device companies is *patient well-being and safety*—in fact this is noted in every pharmaceutical, biotech, and device company's vision or mission statement. As an analogy, let's say a close friend has been hospitalized and the hospital phlebotomist passing by does nothing to help your friend who is groaning in pain. Why? Because it's not her job to help relieve the patient's pain; her job (requirement) is to draw blood. However, the right thing to do would be to try to help the patient by at least informing the nurse to get some pain medication for the patient. Likewise, industry needs to think about what is the right thing to do in the interest of patients, and once we shift the focus and make *patients* the underlying sole purpose for the company's existence, then the answer becomes quite clear. With regard to clinical trial transparency, the right thing to do would be to disclose all clinical trials, both interventional and non-interventional. Most companies have committed to disclose interventional studies, while a few such as GlaxoSmithKline and Eli Lilly and Company have committed to disclose both interventional and non-interventional trials.

The medical publishing profession has recognized this gap and the latest GPP3, published in September 2015, recommends that companies publish both interventional and non-interventional studies in scientific literature. It will be interesting to see how the industry responds to this in coming years, along with the added push from regulators putting more emphasis on real-world evidence data and studies.

Second, the change in mindset is not only needed within the industry, it is equally important and necessary for academic institutions and researchers who conduct research funded by government agencies and charity foundations to uphold the highest ethical standards to ensure that clinical trial results are communicated. Clinical trials would be impossible without patients and research subjects who volunteer their time and, most importantly, themselves to participate in studies in the hopes of advancing science in generating innovative treatments for other patients and future generations. Withholding information also slows down scientific progress.

2. **Set Appropriate Policies and Procedures for Clinical Trial Disclosure and Publications.** Ensure company and institutional policies include a commitment to disclose and publish clinical trials. Employees of companies and academic institutions are

more likely to do what is required of them based on company or institutional policies and procedures as it can have direct impact on their performance appraisal, promotion, and career. Here, senior executives (including CEOs, chief medical officers) can take an active role in establishing policies and commitment to publication that are consistent with *doing the right thing for patients*, and equally important, in making adequate resources available to be able to execute this. I cannot emphasize enough the latter part of the last sentence. Big, lofty commitments will have little to no impact if they are not backed up with enough resources at a grass-roots level to execute and comply with the commitments. Furthermore, company and institutional associates should receive adequate training, along with regular refresher training, to ensure appropriate understanding of the procedures, roles, responsibilities, and accountability.

3. **Allocate Adequate and Appropriate Resources.** Often even with the right intent, one of the major barriers to publication compliance is a lack of or limited resources (i.e., financial and human resources). Getting a paper published in a journal is an extremely long and resource-intensive process, which we have reviewed earlier in Chapters 5 and 6. With regard to industry-sponsored research, the industry spends millions of dollars annually in providing publication support to get industry-sponsored research published, yet it may not be sufficient to cover all studies. Industry publication managers may not have sufficient capacity to manage the workload of publishing all trials, and if this is coupled with limited budgets, the focus will be on studies that are *required* to be published and those that are of interest to the clinical team or investigators. Adequate and appropriate resource allocation, along with use of strategies to minimize costs (e.g., targeting appropriate journals and utilizing publishing platforms to minimize rejections) can help to improve publication success rates.

    With regard to non-industry-funded research, academic clinicians are often also constrained with limited resources and time to dedicate toward publishing. As noted previously, academic clinicians do recognize the value of professional medical writers. However, publication professionals are generally not as commonly employed or utilized directly by academic institutions for non-industry-funded publications. Academic researchers often rely on medical residents or fellows to help with publication writing, and they may not have enough experience with scientific publishing and require further guidance. Support provided by publication professionals is not limited to medical writing and editing; they can also help with management and coordination of draft review, preparation of graphs and figures, and provide guidance on the publication process and requirements, which can be of benefit and value to academic researchers and institutions in meeting clinical trial disclosure and publication obligations. Obtaining adequate funds for professional publication support could be possible through research funders, when requesting the funds for the study itself.

4. **Develop and Execute Clinical Trial Disclosure and Publication Plans.** Study investigators and industry medical associates can proactively prepare integrated clinical trial disclosure and publication plans. A study steering committee (SSC), often established at study initiation, can play an important part in ensuring disclosure and publication of clinical trials. Publication professionals can play an integral role and collaborate with clinical disclosure teams to help ensure that all clinical trials in patients are publicly disclosed and published. The increasing push toward data sharing requirements from regulators, funders, and journal publishers will require companies and institutions to properly plan and coordinate timing and resources to meet necessary requirements. In addition, for publications of clinical trials, inclusion of clinical trial identifier (e.g., NCT number) should become standard and required reporting in all journal and congress publications to help identify and monitor for compliance with publishing clinical research.

Industry-sponsored Phase I trials that involve healthy volunteers may be delayed for publication due to confidentiality, competitive nature or patent filing. However, eventually, these trials could also be published or made public following marketing authorization or patent clearance.

5. **Monitor Compliance with Clinical Trial Disclosure and Publication.** Monitoring progress and status of clinical trial data dissemination on a regular basis is critical and essential in improving compliance with regulations and requirements. With increased scrutiny, many large biopharmaceutical companies have built an infrastructure with dedicated personnel to help monitor compliance with clinical trial disclosure and publication, which is reflected in improved compliance rates for industry-sponsored trials over the years. Some companies opt to hire third-party experts to aid in regular monitoring and to conduct audit inspections. For non-industry-sponsored research, strengthening the academic institution's compliance monitoring efforts could be highly valuable in improving compliance with these obligations.

While clinical trial transparency is one aspect, there are other areas within scientific publishing that also relate to transparency, namely authorship and contributor credit for publications as well as the journal peer-review process.

Authorship and Contributorship for Publications

Authorship on publications has been the established gold standard, a long-standing, and desired credit that academic researchers aim for when conducting clinical research. As discussed previously, this is partly due to the fact that academia rewards researchers with promotions, tenure, and career progression based on publication authorship, number of publications, and the prestige (e.g., journal impact factor) of the journal in which the research is published. With the introduction of the data sharing initiative, academia may introduce additional forms of incentives for researchers to be recognized and rewarded when other researchers use their shared data to publish secondary analyses (Lo and DeMets, *NEJM*, 2016). Lo and DeMets also emphasize the importance of creating a reliable tracking system for secondary datasets and related publications for the researchers to receive such academic recognition. Co-authorship of data generators and data requestors can become common practice for secondary publications in situations where there is full collaboration between the involved parties. Challenges for determining authorship eligibility can arise when the level and scope of involvement from data generators is limited to none in the secondary analysis itself, while it's important to ensure that they are recognized appropriately and prominently on the published articles. Sharing of data from large multicenter trials involving many investigators and secondary analysis involving multiple studies can also pose a challenge for co-authorship.

In order to recognize the efforts and contributions of all stakeholders transparently and accurately, consideration should be given to the current format in which authors and contributors are listed within published articles. For full transparency, readers should be able to view contributors and study sponsor information up front together with the authorship byline. For example, the article's title can be immediately followed by an explicitly labeled list of authors, sponsors, and contributors (including professional medical writers [if involved]) at the top of the paper. To address co-authorship challenges related to secondary manuscripts from sharing data of large multicenter trials, the original study investigators may be acknowledged in the contributor section or by listing them separately as "data generators." Similarly, peer reviewers can be acknowledged for their review effort within the published article either in the contributor section up front or by creation of a separate line of credit—peer review. Academic institutions can elevate the value of peer-review and data generator credit and acknowledgement as potential factors during hiring and promotion decisions. Likewise, PubMed searchability and citation listing should include both author and contributor lists together with the article titles. One of the issues is lack of visibility when acknowledging non-author contributors; elevating the contributor credit line to a more prominent area can also give more visibility.

Success of any new model or approach for granting credit will depend on the acceptance and adoption of the model by academia and researchers to integrate it as part of the academic reward system. Hence, it's important to engage all relevant parties—academic institutions, universities, researchers, journal editors, publication professionals—to review challenges associated with the current model, be open to considering innovative approaches, and carefully consider the practical implications and applications of various approaches prior to setting any new model for research credit.

Journal Peer-Review Process

In an era of transparency, when regulators and policy makers are demanding researchers provide full disclosure of clinical trial results and journal editors are demanding researchers provide full disclosure of publication development, contributors, conflict of interest, and funding source, the peer-review process has also been questioned for insufficient transparency. Medical science journals have been experimenting with various approaches such as double-blind and open peer review, including public disclosure of prepublication history such as identity of peer reviewer and the review report. Some have questioned whether there is a need for peer review at all. Although it can be debated, as it stands now, I believe that peer review is an important part of scientific publishing as it can help ensure scientific credibility (to some degree) of research that is being reported, and help filter out poor-quality research.

As discussed previously, aside from the relatively few journals that have adopted an open access model in some variation, the journal peer-review process remains largely *behind closed doors,* whereby there is no public disclosure of reviewers' identities nor of their review report. Publicly sharing peer-review reports can enable the scientific research community and other health care professionals to know what types of questions or concerns might have been raised by the reviewers (or referees) and how the authors responded along with the revisions to the paper. Such insights can help with the design and planning of future studies and research. Disclosure of peer review can also help build trust and collaboration. In addition, in certain instances, peer reviewers provide extensive comments on manuscript drafts resulting in significant revisions to the paper. Such contribution by peer reviewers remain undisclosed and can be questioned as ghost contribution when their names and input are not published together with the article.

Some journals, primarily open access, have adopted a hybrid or fully open access peer review. Some journals are trying a gradual approach to gauge acceptance of open peer review by giving reviewers the option of being identified and to publish their review together with the published article. The most transparent peer-review process is the one utilized by *F1000Research*, where every stage of peer-review process is publicly available in real-time along with referee names, status of review, and outcome of each referee's review, including the rating and review report once the referee has completed the review.

Increasing experience from journals that have been using open access peer review should hopefully continue to provide further evidence and confidence to other journals in adopting open peer review. One of the concerns raised with open peer review, especially by junior reviewers, is the fear of retaliation for a negative review from their seniors, which could jeopardize their future career. Therefore, junior reviewers may provide favorable review despite having concerns with the manuscript. So far, there is no evidence that open peer review has a negative impact on the quality of reviews. Furthermore, the evolution of workforce over time will comprise largely of millennials and iGen, while the prior generations are retiring. This may also help move further toward open peer review, as the coming generations are accustomed to more open communications and may have fewer concerns about an open peer-review system. In addition, as mentioned previously, academic researchers can be incentivized by receiving credit for their review effort within published articles, and further rewarded as part of their promotion and tenure discussion.

In the future, peer review could perhaps take a completely different form. For example, peer review could be conducted together with artificial intelligence (e.g., IBM Watson or similar technology), which could also help address and alleviate several issues currently facing peer review—short supply of reviewers, timeliness of completing review, etc.

Access to Scientific Information and Data

Access to Patient-Level Data

A cultural shift has taken place in the scientific community, including regulators, academia, the industry, and journal editors, in recognizing the significant value of transparency and access to patient-level data. This will further evolve with the focus now shifting to how data should and can be shared. Most importantly, while access to patient-level data is desired, it should not be at the cost of infringing rights to privacy. Projects such as Vivli show that the thinking is in the right direction—collaboration, utilization of existing tools and databases, leveraging prior experiences on processes and operationalization, and addressing the needs of the entire spectrum of researchers (i.e., those from large and small academic institutions to independent researchers). Given the importance of this topic, an entire chapter was dedicated to this and details of what data sharing could look like in the near future has already been described in Chapter 8.

As mentioned previously, it is often the fear of the unknown that holds us back from trying something new and different than what we are accustomed to. The concept of data sharing is relatively new, and thus far, data utilization through existing data-sharing platforms has been limited (Navar et al., *JAMA*, 2016). As researchers gain more experience and confidence and see the benefits of data sharing through personal experience, it should become part of standard practice within scientific research. Data sharing requirements and activities should become integrated with publication planning through coordinated efforts in ensuring timely data availability.

New technological advances should also help generate increased collaborations across various sectors. Sharing and exchange of *clinical trial* data is just one area; in the current era of digitalization, wearable technology, and artificial intelligence, data can be obtained and used for research and discovery from many different sources. Possibilities are endless; the big-data revolution is already underway in health care. Technology giants such as Google, Apple, and Microsoft have already started to tap into health care and forge collaborations with health care companies that will create new business models and transform health care. We are moving away from reactive patient care to health care that is proactive, preventative, and personalized. Overall, I believe that the sharing and exchange of data will play an instrumental role in disruption of health care for the good of society.

Access to Journal Publications

In the current culture of freely available apps and tools, there has also been a welcome change in access to published literature. Nearly all journals now have an online presence and many have also digitized the historical archives of articles that were previously published prior to the online platforms. Thousands of scholarly articles can also be accessed freely from open access journals and online repositories. While this is significant progress in providing access to literature, many of the traditional journals remain predominantly subscription based with open access only to select articles published in their journals (referred to as a hybrid open access model). Readers who do not have a subscription to the journal may end up paying upwards of $30 to $40 per article for non-open access articles. We can now get an entire e-book at a fraction of the cost of what we have to pay for a single article. Journal publishers also tend to push for an all or none approach when offering subscription deals to institutions; where overall, it would be cheaper to pay a subscription for the entire portfolio of journals owned by the publishers than to purchase subscriptions for a subset of journals. However, the discounts offered through bundling have also helped academic institutions in light of budget pressures.

The reasons for relatively slow progress in standardizing open access for all journals can be attributed to several interdependent factors:

- The majority of traditional journals are owned by for-profit journal publishers, and subscription-based model is likely to provide greater profit margins than open access.

- Academic institutions and other organizations continue to make investment in journal subscription fees, which could instead be used for more research and education. Although academic institutions have been under pressure to reduce spending and have done so, academia has not been able to move completely away from subscription-based journals. Therefore, as long as there are institutions that are willing to pay for journal subscriptions, journal publishers will be less likely to change their business model.

• Academia continues to recognize and reward researchers based on journal prestige and impact factor, hence researchers' desire and aim to publish in traditional journals that tend to have higher impact factors, which generally follow subscription-based model. In addition, since the subscription fees are paid by the institutional libraries, individual researchers do not see the costs directly.

How might this evolve further? One major factor that has and will likely continue to help move further toward increased access to scientific publications is the mandate imposed by funders (e.g., NIH, the Bill and Melinda Gates Foundation, and the Wellcome Trust) for peer-reviewed publications to have unrestricted open access. However, in order to see a significant shift, the mandate needs to be considered by the wider scientific community, including academic institutions and the industry. If academic institutions and pharmaceutical industry organizations such as PhRMA and EFPIA agreed to support and publish their research exclusively in an open access model, then the journals would be forced to relook at the current traditional business model and adopt more innovative solutions that include open access as a standard, not as an option! Secondly, with further increases in budget pressures and the availability of open repositories, academic institutions could stop institutional journal subscriptions altogether (which would be an extreme situation). Now, if the burden of cost is transferred to the individual researchers, they will be more likely to opt for publishing in open access journals, so they do not have to pay for their or others' published articles. Thirdly, the academic reward system could reward researchers for publishing in open access journals rather than based on journal prestige and impact factor, which could also help with practice and behavior change. Lastly, the need for printed copies of journals will likely diminish over time. The next generations of scientists and health care professionals who have grown up with computers and handheld smart devices are more likely to prefer an electronic format, and if there is any interest in having a printed copy, they can print the article on their own printer.

As noted before, the millennials and iGen, who are conditioned with a lifestyle of choice, on-demand purchasing behavior (buying only what they want/need when they want), and having access to freely available information and tools, are not likely to be willing to buy subscription for an entire journal nor pay the customary $30-40 per article. This may require a critical look at long-term consequences of continuing with the subscription-based model. As an analogy, when Netflix first come out, Blockbuster had similar conservative beliefs for movie rentals and believed that consumers would continue to rent movies as they did in the past. Now Blockbuster does not exist, and Netflix is worth $40 billion and movie streaming is the norm. Leveraging technological advances while embracing newer business models that address changing user trends and behavior will be important for journal publishers to continue to succeed.

One-Stop-Shop Access to Publicly Available Clinical Trial Information

As discussed previously, there is a tremendous wealth of scientific information already available in the public domain. However, this information is currently fragmented. For example, published clinical trial results can be found on individual, segregated platforms and repositories such as PubMed for peer-reviewed journal articles; PubMed Central and Europe PubMed Central for free full-text articles; individual industry-sponsored websites for clinical study reports and synopses; regulatory repositories such as clinicaltrials.gov, EMA websites, and other national registries for trial protocols, study reports, and summaries; university and academic institutional repositories; scientific congress websites for published abstracts, posters, and oral presentations; and other sources such as ResearchGate, etc. The ability to search and access all publicly available documents, reports, and articles in a single portal or platform would significantly improve efficiency and productivity for information gathering and provide a more complete picture of available information within the public domain. The availability of such a platform would benefit the whole health care enterprise as well as patients while enriching the entire international scientific and medical community with available knowledge.

OpenTrials (www.opentrials.net), recently launched in October 2016, is a centralized, one-stop-shop, freely accessible online portal that pulls clinical trial information from multiple sources—registries (including regional and national), company databases, PubMed, and publicly available regulatory documents. The organizers of this initiative also aim to gather and archive additional documents related to clinical trials based on targeted crowdsourcing (Goldacre and Gray, *Trials*, 2016). This platform could become an useful tool to many stakeholders:

- Researchers or investigators could learn about range of clinical trials ongoing or completed in specific patient populations, disease area, and interventions
- Practicing clinicians and healthcare professionals interested in learning about available clinical trial data and evidence to help with better patient care
- Patients interested in participating in clinical trials that are actively enrolling in their geographical region

As this platform is likely to be utilized by many stakeholders, including patients, it is extremely critical that the information available on this portal is accurate and consistent with the information archived in multiple source databases. Planned crowdsourcing to collect additional documents and information for archival on this platform can pose some challenges such as questionable accuracy and reliability of information, maintenance of up-to-date documents, credibility of the person who uploads the documents, and appropriate permissions from trial sponsors or owners of the documents. Over time with more experience, the portal's benefits, usefulness, functionality, and limitations should become more evident.

Alternative Article-Level Metrics

In my opinion, alternative article-level metrics are the future and the most appropriate way of measuring the value of research. Most of the current utility, research and discussions on the use of altmetrics has been in relation with citation analysis and journal impact factor. Earlier studies suggested social media attention to be early predictors of citation rate (Thelwall et al, *PLoS ONE*, 2013; Knight, *Transplantation*, 2014), while more recent studies have shown a lack of or weak correlation between Altmetric scores and number of citations (Peters et al, *Scientometrics*, 2016; Barbic et al, *Academic Emergency Medicine*, 2016). Skobe et al. assessed the Altmetric score of 408 company-sponsored manuscripts published between January 1, 2013, and December 31, 2014, and reported that eight out of ten articles with the highest mentions were published in the *NEJM* (Abstracts from the twelfth annual meeting of ISMPP, *CMRO*, 2016). As expected, the highest number of mentions and activity was two to three weeks from publication when the news of the results are relatively new and of interest. The authors did not mention if this research allowed them to identify potentially lower tier journals that might provide higher mentions or Altmetric scores. *NEJM* generally accepts publications that are based on key pivotal studies or data that are often groundbreaking and which the medical community may be anticipating or have increased interest in. This, coupled with *NEJM*'s social media presence and promotion, makes it not surprising that the articles published in this journal received higher Altmetric scores indicating significant social and peer attention. There is an inherent selection bias introduced by the journal editorial and peer-review process for articles of interest, which can result in higher Altmetric scores for articles published in high-ranking journals, while those published in lower tier journals and/or journals without sufficient social media presence may likely achieve lower scores.

Given the current knowledge and understanding of altmetrics, they can be considered as a complement to the traditional metrics. The true value of alternative article-level metrics is more likely to be appreciated in the absence of journal selection bias. Self-published articles and datasets on an online peer-review publishing platform would be a suitable forum for article-level metrics. Furthermore, altmetrics are relatively new measures and more data are needed on their use and interpretation.

Speed to Publish

Of all the areas relevant to scientific publishing, the time it takes to publish a paper in a peer-reviewed journal has not improved much over the years. The publication time that the journals promote in their marketing materials are specifically for the review time at their journal. However, a manuscript lifetime can be much longer due to the numerous review–rejection cycles it undergoes before finally getting published. A manuscript's prepublication lifetime can be difficult to ascertain as it is not common practice for academic researchers to systematically keep track of timing of multiple journal submissions, rejections, and resubmissions. Moreover, based on experience, publication time can range from as quickly as six to nine months to more than two to three years from the time of first journal submission. Among 635 NIH-funded trials, it was reported to take a median of 23 months (range, 14 to 36 months) from the time of trial completion for 68% of the trials to be published in a peer-reviewed journal (Ross et al., *BMJ*, 2011). Median time to publication from trial completion date ranged from 15 to 31 months across 51 academic medical centers involving 4,347 trials (Chen et al., *BMJ*, 2016). Siler et al. (*PNAS*, 2015) also noted that, in general, the papers would eventually get published within two years from the time of the first journal submission. Pharmaceutical companies utilize electronic publication tracking tool, which allows companies to monitor the progress of publication drafts and review process. In an analysis of 410 published papers of company-sponsored research, Whann et al. (Abstracts from the ninth annual meeting of ISMPP, *CMRO*, 2013) noted that publication of papers in second journal submission was extended by five months compared to those published in the first target journal. In another study of 41 industry-sponsored research publications, Truss et al. (Abstracts from the twelfth annual meeting of ISMPP, *CMRO*, 2016) noted an increase in the cumulative time to acceptance after each rejection and resubmission of manuscript to a new journal.

It's no surprise that the international consortium of academic researchers proposed a window of a minimum of two years post-publication for access to data. Researchers are aware of how long it takes to publish and hence want to have enough time to publish on their own before sharing the data with others. Although I understand their dilemma, the solution to that issue should not be delaying data sharing, but to consider alternate approaches to speed up publishing. Researchers have often been implicated in delaying preparation of a manuscript for publication, however, the long publication time can also be partly due to inefficiencies in the journal publication process and requirements, namely journal-specific peer review and manuscript format for submission, and sequential submission of manuscripts—one journal at a time, to name a few.

Standardization of Manuscript Submission Format

Standardizing manuscript submission format is a low hanging fruit that, I believe and hope, can be addressed quickly. As mentioned previously, in the process of submitting a paper from journal to journal due to rejection, researchers and authors end up spending significant time in reformatting the manuscript. Additionally, based on my experience within pharmaceutical industry, collectively, it can cost millions of dollars annually just for manuscript reformatting support for industry-sponsored research publications. I think most people would agree that this time and money can be better utilized for more productive activities such as more research and working on publishing unpublished data, to name just two. Having a standardized format for manuscript submission across all journals could improve publication time substantially while making the process a lot more efficient.

Interestingly, this is not the first time such a request or suggestion has been made. In 1968, Augusta Litwer, a secretary to nephrologist Belding Scribner, grew tired of retyping his papers to change the format of the manuscript when a paper was rejected by a journal and needed to be reformatted to the other journal's requirements. Litwer wrote a letter to the editors of *Annals of Internal Medicine*, the *JAMA*, and the *NEJM* inquiring why they could not have a standard format for references. The editors responded positively, which eventually led to the birth of what we now know as ICMJE (Huth and Case, *Science Editor*, 2004). The first and early versions of ICMJE's Uniform Requirements for Manuscripts Submitted to Biomedical Journals (URM), in 1978, included recommendations on standardization of manuscript format and preparation across journals. The guidelines have since evolved to address the ethical issues related to scientific publishing. So now, nearly 50 years later, we have come full circle and are requesting a standardized manuscript format again.

Preference of manuscript format, style, and referencing are cosmetic attributes that allow the journals to maintain a unique identity. If the journals wish to maintain their unique format and requirements, they can develop or use software to convert the standardized manuscript into whatever format they choose to have using their own resources, once the paper is accepted for publication by the journal. Alternatively, it would be reasonable to require the authors to format the manuscript per journal requirements only after it is accepted for publication by the journal. This would be much more efficient since the authors would need to prepare the manuscript per journal requirements only once as opposed to repeated reformatting for every resubmission.

ICMJE can help drive the change by creating and implementing a standardized manuscript format for submission, including word count limit, tables, figures, referencing, etc., for its member journals. Other non-member journals would likely follow suit, which can ultimately help speed up time to publication while reducing or eliminating the burden and cost associated with manuscript reformatting.

Centralized Peer Review

Peer review is generally journal specific. A paper can undergo peer review at one journal and get rejected, and it will undergo another round of peer review at the next journal. In many specialties, the reviewer pool is quite small and the same reviewer may end up receiving the manuscript for review from journal B, which he/she may have already reviewed for journal A. Thus, creating inefficiency and consuming additional time and effort of the reviewers for manuscript review. Journal peer review for publications can be considered analogous to review and approval by the institutional review board (IRB) for clinical trials. An institutional or medical center's IRB must approve the study protocol before the site agrees to participate in a clinical trial. The traditional approach to IRB review for clinical trials is that each participating site requires its own IRB review and approval, which often results in delays in study start-up and patient enrollment. Within the specialty of oncology, numerous research networks have been established, and a key advantage they offer is centralized IRB review and approval. Each network includes multiple clinical centers or sites within a region or across the country. When working with a research network, the sponsor needs to get the protocol reviewed and approved by the centralized IRB, which opens up multiple sites within the network for patient enrollment. Such networks, available in the United States as well as Europe, have facilitated much faster patient enrollment for oncology trials.

Journal publishers and editors can take this as best practice and apply a similar concept of centralized peer review for publications as well. The main purpose of peer review is to get critical expert review of the paper, and getting it done through centralized peer review should address this objective. The cascade and portable peer-review concepts discussed earlier are along the same lines as this; however, creating a centralized peer-review system that is recognized by many or all journals within each category or specialty would be taking it one step further. Centralized peer review should help expedite the publication process by alleviating multiple rounds of peer review.

Innovation in Scientific Publishing Platforms

Nearly all sectors have taken advantage of advanced technology and digitalization to accelerate and enhance products, systems, and services with out-of-the-box innovative solutions. For example, telecommunications has transformed from the use of hardwired desk phones to handheld phones to mobile phones to smartphones to wearable devices in less than three decades. In comparison, to their credit, traditional journals have adopted the use of online submission platform (as opposed to submitting via email or by mail), social media, metrics for articles that are published online, and hybrid open access as some of the enhancements to help with the publishing process and experience. Aside from the online manuscript submission platform, there are no other enterprise-wide changes to help speed up the publishing process. As mentioned, the cascading and portable peer review are additional variations that some journals are experimenting with in an effort to improve efficiency; these approaches are still in the infancy stage and not yet adopted as a standard across all life sciences journals.

There have been a few pioneers who have created some exciting online platforms and alternative approaches within the scientific publishing landscape. Let's take a look at some of the available online platforms.

Repositories With or Without Social Network

- **ResearchGate** (www.researchgate.net) is a free online platform that boasts about three million members from academia, mostly in biological sciences and medicine, and allows researchers to publish and share instantaneously, receive feedback from peers, ask questions for feedback, and create opportunity for research collaboration. It's a fusion of publishing and social network and promotes itself as a social networking site. Some of the social networking aspects include the ability for members to create profiles, like and follow researchers and their publications, endorse the skills of others, bookmark favorites, and share news and posts. From a publishing standpoint, it does not provide a systematic, formalized peer-review process in order to generate a citable peer-reviewed reference. For instance, members can upload their draft manuscript to obtain feedback from network members, which can serve as a preliminary, informal review and feedback to improve the paper before it is submitted to a journal (which is often referred to as preprint or prepublication review). Hence, although this provides the benefit of a review prior to journal submission, the articles will not be indexed in PubMed and researchers would still be required to submit their paper to a journal as usual to make their paper a peer-reviewed, referenceable article. Secondly, the site is popular for researchers to upload their papers that have already been accepted and published by a journal, and hence it acts as a repository to provide free access to articles that are already published, and these articles are indexed by and accessible through Google Scholar. ResearchGate automatically matches the uploaded article against the SHERPA RoMEO database of journals to ensure that the uploaded articles are legal and do not infringe on the journal's copyright policy. Nevertheless, uploading of published articles to ResearchGate may not address the researchers' obligations related to funder policy. For example, articles may still need to be posted on open-access databases according to a funder's policy (e.g., PubMed Central).

On the other hand, the site provides its members the convenience of searching and accessing publications from many free online databases such as PubMed and arXiv, as well as certain university websites and repositories.

- **Academia.edu** is similar to ResearchGate, with over two and a half million academic researchers who are able to share their publication drafts or final journal publication with its network members, receive feedback, connect with other researchers, and gauge popularity of their work through site metrics. Similar to ResearchGate, again, this platform also does not provide the possibility of generating a peer-reviewed referenceable article, and researchers still need to go through the traditional journal submission and acceptance for publication.

- **Mendeley** (www.mendeley.com) is another reference social network site with over five million member researchers who can share their published articles, connect with fellow researchers, follow peers to view their research, measure article performance (downloads, citations), and identify collaborative opportunities. The unique offering of this platform is the ability to generate citations and bibliographies for a new publication draft, and hence it promotes itself as a free reference manager. In addition, it allows readers to read and annotate published articles for future reference. Again, this platform does not offer the possibility of generating a peer-reviewed article.

- **Social Science Research Network** (SSRN; www.ssrn.com) is a platform that provides a central repository or e-library for publications generated by members of multiple research networks across social sciences by partnering with the publishers of journals relevant to various social sciences specialties. There is no social network functionality on the site, and in fact, the readers need to pay for access to articles according to each journal's policy. In May 2016, SSRN was bought by Elsevier.

Preprint Archive

- **ArXiv** (www.arxiv.org), developed by Cornell University, is a popular, established platform for research covering non-medical topics such as statistics, physics, computer science, mathematics, finance, and quantitative biology. It was launched in 1991, and its existence served as precedence and best practice to help with the open access movement overall. It is referred to as a repository of preprints since the papers archived on the site are not considered peer-reviewed. It allows researchers to self-publish their manuscript draft on the site and receive feedback from their peers. Articles that are offensive and/or have non-scientific content are rejected. Researchers can choose to further submit their paper to a journal along with the comments received on arXiv to demonstrate how the paper was improved.

  More recently, in 2013, Cold Spring Harbor Laboratory introduced **bioRxiv** (www.biorxiv.org), a free online preprint platform similar to arXiv but specifically for life sciences, including clinical trials. The platform also allows researchers to have direct transfer of their manuscript and related files (e.g., reviewer report) from bioRxiv to a participating journal's submission system.

- **PeerJ PrePrint** (www.peerj.com). Similar to bioRxiv, this platform is designed for researchers in biological and life sciences to obtain preliminary feedback on publication drafts, which can then be submitted formally to the publisher's open access journal (*PeerJ*) or another journal for the traditional journal review process.

Self-Publishing E-Book

- **Amazon Kindle** self-publishing platform (https://kdp.amazon.com/) has revolutionized and disrupted the traditional publishing model for non-scholarly publications, which interestingly has a similar track record of significantly long publication time as we have seen for scholarly publications. Currently, Amazon's self-publishing platform is not set up for publishing scholarly scientific research, but it would be interesting to see what would happen if they decided to tap into this space!

- **Lulu** and **Smashwords** –These two platforms are primarily for self-publishing scientific e-books and textbooks. Hence the scope of these platforms is not relevant to publishing peer-reviewed research articles.

Self-Publishing Scholarly Articles

- *F1000Research* (www.f1000research.com) is a scientific self-publishing platform with a final output of peer-reviewed article indexed on PubMed, PubMed Central, Europe PMC, Scopus, Chemical Abstracts Service, British Library, CrossRef, directory of open access journals (DOAJ), Google Scholar, and Embase. It offers prompt publication of a pre-peer-review version within seven days of submission. The seven-day delay, according to their website, appears to be for initial editorial check, plagiarism check, etc. If the quality of writing is judged to be unpublishable, then the article could get rejected outright before it is published online to proceed with peer review. Although the publisher has an advisory board of over 1,000 experts, it does not follow the traditional editorial board format. Peer reviewers are invited by *F1000Research*; however, the selection of reviewers is based on author recommendations, which can result in potential reviewer bias. The peer review process is open access, including reviewer names, review reports, and status of review. However, the peer-review process does not seem to be monitored for completeness or rigor of review. Platform offers the ability to cite the paper during peer review (prior to the final indexed, peer-reviewed version) based on assigned DOI number. The site also retains article that were unapproved by peer reviewers or with incomplete reviews. In addition, the articles may get indexed even if two reviewers *approve with reservations*. Hence, although the platform offers open peer review, it does not appear to have a rigorous process for ensuring quality of review. The platform provides a range of article-level metrics such as number of views, downloads, and recommendations by members of the F1000 Faculty; however, it does not carry the traditional impact factor.

Other Tools

- **Publons** (www.publons.com) allows reviewers to archive, share, and track the peer-review critiques on a single platform. Researchers can generate a bibliography of completed peer-review reports to highlight their contributions in their funding applications and for job promotion. It also allows journal editors to track the manuscripts managed by the individual editor, along with the peer reviewers used for those manuscripts.

- **ORCID** allows researchers to register and generate unique personal digital identifier that can be used when applying for research grants, submitting publications, and any research workflow to ensure attribution of credit for the work.

- **Manuscript submission platform.** Most journal editorial offices utilize an online platform to manage manuscript submission, peer review, and publication process. Journals have the option of using ScholarOne, Editorial Manager, eJournalPress, Produxion Manager, Bench>Press, or Open Journal Systems.

- **Publication management system.** There are several publication management systems available such as Datavision, PubStrat, and PubsHub. These systems are customizable and commonly used by biopharmaceutical companies to help with publication planning, tracking, and managing draft review workflows, including archival and documentation of author review and approval of drafts. Regular updating and monitoring of the system is required to ensure accuracy of information, which also allows to maintain live record of the publication plans.

In summary, there are many innovative solutions and platforms that have become available over the last decade. The majority appear to be repositories of published articles, while others offer a system for self-publishing, mainly as a preprint alternative. The availability of online platforms, such as manuscript submission portals and publication management systems, have provided innovative solutions to improve efficiencies in publication process.

The uptake of scientific self-publishing systems has been sluggish within life sciences journal with majority of its use for biological preclinical research. Traditional journals and systems seem to predominate for clinical research papers. Academic reward system at institutions have also remained dependent on traditional journals; there is no incentive for researchers to publish quicker or in open access journals. Changes in behavior often require significant effort and time, especially when it involves long-standing, established systems. As mentioned previously, most are inclined to do what they are *required to do*. Based on past patterns, it appears that unless there is an incentive, regulation, or policy that requires publishing clinical research within a specific period of time, the adoption and transition to self-publishing and open access systems may be quite slow.

The available online platforms can serve as an opportunity to gain key insights and learnings for what works and how future platforms can be improved. There is a significant opportunity for traditional journals to adopt and use the technology and models that have been tried and tested. Traditional journals can coexist with the newer publishing options, and the process of conducting peer review can be modified while leveraging currently available technology.

Closing Remarks

Scientific publishing will remain the cornerstone of research communication. It's the window through which researchers showcase their research and exchange ideas and opinions. Scientific integrity, transparency, access, and speed are the key critical elements that will continue to shape the future of the scientific publishing system. The cultural shift focusing on transparency and access has made tremendous in roads and paved the path for further reform in the future. Let's celebrate our achievements thus far and continue with *the right thing to do* for patients and society! Let's keep the momentum going for recoding our scientific publishing ecosystem while raising the bar for efficiency, transparency, access, and speed during the transformation.

I would like to leave with the following parting thought. It's a saying that I was introduced to by my teacher of Hindi in grammar school back in India:

Gnyan baatne se badhta hai. (in Hindi)
Literal translation of this is...*Knowledge grows [expands] by sharing.*

I would sincerely like to thank you for reading this book. I hope you enjoyed it and that it stimulates new ideas and ways to further recode scientific publishing!

For assistance with scientific publications, or if you wish to share your comments/suggestions, please visit us at <u>www.publicationpracticecounsel.com</u>.

I would love to hear from you!

Glossary

Abstract: This is a type of publication that is submitted to and published by a scientific or medical congress. It is intended to provide brief summary of the research and its results, typically, in around 500 words. The congress can either reject or accept an abstract for presentation in one of two ways—poster or oral presentation. Journal articles also often include abstracts as a summary of the paper.

Academic institutions: Within this book, academic institutions refer to universities, medical schools, and medical centers dedicated to education and research.

Academic research organizations (ARO): See description for 'Contract research organizations'.

AMWA: American Medical Writers Association (www.amwa.org), established in 1940, is a professional organization for writers, editors, and publication professionals with approximately 5,000 members.

Author: Broadly, it is defined as "a writer of a book, article, or report." In scientific publishing, an author is generally defined according to the International Committee of Medical Journal Editors (ICMJE)'s authorship guidelines (www.icmje.org). Authors of scientific research publications are typically those who have participated and contributed to study design, conduct, analyses and/or interpretation of results, and have met all four ICMJE authorship criteria.

BMJ: British Medical Journal

CMRO: Current Medical Research and Opinion

Contract research organizations (CRO): CROs support industry with outsourced services related to research, including product development, biologic assay development, preclinical research, clinical research, clinical trial management, and pharmacovigilance. CROs can also support research institutions, universities, government organizations, and foundations that support research. Independent CROs have been utilized conventionally, which range from large international organizations to smaller boutique companies. More recently, university-based **academic research organizations (ARO)** have emerged and gaining popularity, which offer similar outsourced services to the industry for clinical research and trial management as CROs while ensuring independent academic oversight of the research and data analyses for greater assurance of integrity (e.g., journal and regulatory requirements for verification of statistical analyses, study conclusions, etc.).

Copyright: It is defined as "exclusive legal right, given to an originator or an assignee to print, publish, perform, film, or record literary, artistic, or musical material, and to authorize others to do the same." Types of copyright licenses according to Creative Commons Corporation (www. creativecommons.org):

> *CC-BY* allows others to distribute, remix, tweak, and build upon one's work, including commercial purposes, as long as they credit the individual or entity for the original creation. This allows maximum dissemination and use of licensed materials.

> *CC BY-SA* allows others to remix, tweak, and build upon one's work, including commercially, as long as they credit the originator and license their new creations under identical terms. This is the license used by Wikipedia, and is recommended for materials that incorporate content from Wikipedia and similarly licensed projects.

CC BY-NC allows others to remix, tweak, and build upon one's work non-commercially. Although the new works must also acknowledge the originator and be non-commercial, they don't have to license their derivative works on the same terms. If the material is used for commercial purposes, then it will require the user to obtain permissions and pay copyright fees as applicable.

CC BY-NC-SA allows others to remix, tweak, and build upon one's work non-commercially, as long as they credit the originator and license their new creations under the identical terms. If the material is used for commercial purposes, then it will require the user to obtain permissions and pay copyright fees as applicable.

CC BY-NC-ND allows others to download one's works and share with others as long as they credit the originator, but the work cannot be changed in any way or used commercially. This is the most restrictive of the six licenses. If used commercially, the user is required to obtain permissions and pay copyright fees as applicable.

DOI: Digital object identifier is a type of identifier used to uniquely identify electronic documents (objects), namely journal articles and datasets. Launched in 2000, the DOI system is managed by the International DOI Foundation.

EFPIA: European Federation of Pharmaceutical Industries and Associations (www.efpia.eu) is a trade association including membership of 33 national pharmaceutical associations and 41 leading pharmaceutical companies within Europe. EFPIA provides advocacy for the research and development efforts and a voice for 1,900 pharmaceutical companies within Europe. The association has a number of position statements, codes, and guidelines on ethical practices for pharmaceutical companies.

EMWA: European Medical Writers Association (www.emwa.org), founded in 1989, is a network of publication and medical communication professionals that represent Europe with over 1,000 members.

Good Publication Practice (GPP): "Good Publication Practice for Communicating Company-Sponsored Medical Research: GPP3" is the current and third edition (2015) of the guidelines for planning and development of publications of research sponsored by pharmaceutical, biotechnology, medical device, and diagnostics companies. The guidelines were first published in 2003 and its aim is to provide standards and ensure that "clinical trials sponsored by pharmaceutical companies are published in a responsible and ethical manner." GPP3 can be accessed at http://annals. org/aim/article/2424869/good-publication-practice-communicating-company-sponsored-medical-research-gpp3

ICMJE: International Committee of Medical Journal Editors is comprised of the US National Library of Medicine, the World Association of Medical Editors, and 11 member journals including *Annals of Internal Medicine, British Medical Journal, Chinese Medical Journal, Deutsches Ärzteblatt (German Medical Journal), Ethiopian Journal of Health Sciences, JAMA (Journal of the American Medical Association), New England Journal of Medicine, New Zealand Medical Journal, PLOS Medicine, The Lancet, and Ugeskrift for Laeger (Danish Medical Journal).* The committee publishes the most widely recognized and utilized guidelines "Recommendations for the Conduct, Reporting, Editing and Publication of Scholarly Work in Medical Journals". Further details can be found at www.icmje.org.

ICMJE authorship criteria: As part of the guidelines, ICMJE has defined specific criteria for authorship of publications. This helps to ensure consistency in defining an author versus other non-author contributors. In addition to the below criteria, the guidelines also provide further guidance on the role and responsibilities of an author. In order to be listed on publication author byline, all authors must meet the following four authorship criteria:

1. Substantial contributions to the conception or design of the work; or the acquisition, analysis, or interpretation of data for the work; AND

2. Drafting the work or revising it critically for important intellectual content; AND

3. Final approval of the version to be published; AND
4. Agreement to be accountable for all aspects of the work in ensuring that questions related to the accuracy or integrity of any part of the work are appropriately investigated and resolved

IFPMA: International Federation of Pharmaceutical Manufacturers and Associations

Industry: Within this book, pharmaceutical, biotechnology, and medical device companies are collectively referred to as 'industry'.

Industry-sponsored study: These are studies in which the industry plays the role of study sponsor (also see 'Study sponsor' for more information), while the academic researchers conduct the study as an investigator.

Investigator initiated trial (IIT): IITs are clinical trials that are initiated and conducted by an individual (researcher) or academic institution, who plays the role of both a study sponsor and an investigator. The study may be financially funded by industry, government agencies or private organizations. Please refer to Chapter 2 for more details.

ISMPP: The International Society for Medical Publication Professionals (www.ismpp.org) is a non-profit organization with more than 1,500 members representing publication professionals from pharmaceutical, biotechnology, and medical device companies; medical communication agencies; journal publishers and editors; and professional medical writers. ISMPP provides advocacy for medical publication profession, helps set ethical standards for medical publications, and leads key initiatives such as the Good Publication Practice (GPP; see further description for 'Good Publication Practice').

JAMA: Journal of the American Medical Association

Journal impact factor (JIF): JIF is a metric calculated based on number of citations received by articles published in that journal during the two preceding years, divided by total number of articles published in that journal during the same time frame. Despite much criticism on its validity as a measure of importance, it is the most widely recognized metric among academic researchers, institutions, and the industry, and continues to be used as one of the criteria for target journal selection.

JPMA: Japan Pharmaceutical Manufacturers Association

Manuscript: This is a draft of an article that an author submits to a journal for publication. It is a detailed report of the research, including an abstract summary, introduction/background, study methods, results, discussion and conclusion.

Medical writer: Professional medical writers are those who assist in writing scientific documents while ensuring compliance with regulatory, journal, and other guidelines. Within the industry, medical writing can be categorized as regulatory writing or educational writing. Scientific publishing entails educational medical writing.

Meta-analysis: See description for 'Systematic Review'.

MPIP: Medical Publishing Insights and Practices initiative (www.mpip-initiative.org) involves a group of members of pharmaceutical industry and ISMPP. The initiative's aims are to "raise standards in medical publications," and to "elevate trust, transparency and integrity in reporting of industry-sponsored research." The initiative has developed and published several guidelines on key topics such as transparency of clinical research, authorship, and adverse event reporting, in partnership with editors of prominent biomedical journals, including *Annals of Internal Medicine, Blood, British Medical Journal, Canadian Medical Association Journal, European Respiratory Journal, Journal of the American Medical Association, Journal of Clinical Oncology, Journal of the American College of Cardiology, Nature Medicine, Neurology, The New England Journal of Medicine,* and *The Lancet.*

NEJM: New England Journal of Medicine

Oral presentation: This is a type of publication that is presented at podium at a scientific or medical congress, and holds higher significance than poster. There are a limited number of oral presentations at a congress compared to poster presentations, and hence acceptance as oral presentation is generally reserved for pivotal, groundbreaking research and considered more prestigious.

ORCID: A non-profit organization that allows researchers to register and generate unique personal digital identifier that can be used when applying for research grants, submitting publications, and any research workflow to ensure attribution of credit for the work.

Peer review: This is a process that involves evaluation of the scientific or professional work by others (peers) working in the same field. Within scientific publishing, peer review refers to evaluation of the draft manuscript submitted to a journal in consideration for publication.

PhRMA: The Pharmaceutical Research and Manufacturers of America (www.phrma.org) is a membership-based trade organization comprised of biopharmaceutical research companies of the United States. It provides advocacy within the United States and worldwide for the research and development efforts of biopharmaceutical companies. The member companies agree to abide by the PhRMA policies, codes, and guidelines for ethical practices in areas such as research and development, promotional or marketing practices, and engagement with healthcare professionals.

Poster: This is a type of publication that is presented at a scientific or medical congress. Posters provide more details of the research and results compared to the abstract. It generally follows congress-specific template, and includes a mixture of text, tables, figures, and illustrations. Scientific posters are typically presented on a poster board; while they can also be presented in an electronic format on a monitor (which allows more interactivity by attendees).

Primary article: This is a type of journal publication that reports primary endpoints and key secondary endpoint analyses of the research. Generally, there is one primary article for a study, which can be followed by multiple secondary articles.

Publication: Broadly, it is defined as "communication of information to public." In this book, it specifically refers to communication of scientific research and results. Publications can be grouped into 2 major categories—congress and journal. Congress publications include abstracts, posters, and oral presentations. Journal publications include primary articles, secondary articles, review articles, case studies, letters to the editor, brief communications/editorial, etc.

Publication manager: Within the industry, a publication manager is the string that brings together the key stakeholders—study investigators, sponsor study team (medical and statistics), medical writer (if applicable), and other experts relevant for the publication. They provide expertise in publication planning and management of time lines and resources for publication development. They also help ensure that publications comply with relevant guidelines (e.g., journal, ICMJE, GPP, etc.).

Review article: This is a type of publication that summarizes published literature on a particular topic. These articles generally express opinion and implications of published data by the subject matter expert. They are also referred to as opinion papers or narrative reviews. Published review articles can be solicited or unsolicited by journals.

Scientific congress: It is a formal meeting of medical or health care professional societies to exchange, discuss and present new research findings and key topics of interest. The congress can be specific to therapeutic area, health care profession or geographical region (e.g., international, regional, or national).

Scientific journal: This is a periodical publication that publishes original research articles, review articles, case reports, editorials, letters to the editor, etc. They can be in traditional subscription model, open access, or hybrid (part subscription, part open access offered as an option to authors). Journals can be in print format, online electronic format, or both.

Secondary article: This is a type of journal publication that reports secondary analyses of a study. Articles reporting additional subset, ad-hoc, or exploratory analyses are also considered secondary articles. A study can have multiple secondary articles.

Study sponsor: Study sponsor is a regulatory terminology. It is defined as an individual, institution, or organization that initiates, manages, and/or finances the research. The actual study investigation may be conducted by an investigator instead of the sponsor.

Systematic review: This type of review involves *a priori* detailed plan of search strategy for the publications, defined inclusion and exclusion criteria for publications that will be selected, and planned methodology for assessing the publications. This systematic approach is intended to reduce selection bias. A systematic review may or may not include **meta-analysis**, which is further assessment of research findings across multiple studies through statistical analyses (e.g., quantitative estimate or summary effect size).

Suggested Reading

Akerlof GA. Writing the "The Market for 'Lemons'": a personal and interpretive essay. November 2003. Accessed: http://www.nobelprize. org/nobel_prizes/economic-sciences/laureates/2001/akerlof-article. html

Anderson ML, Chiswell K, Peterson ED, et al. Compliance with results reporting at ClinicalTrials.gov. *N Engl J Med* 2015;372:1031-1039.

Barbic D, Tubman M, Lam H, et al. An analysis of altmetrics in emergency medicine. *Acad Emerg Med* 2016;23(3):251-268.

Battisti WP, Wager E, Baltzer L, et al. Good publication practice for communicating company-sponsored medical research: GPP3. *Ann Intern Med* 2015; 163:461-464.

Bierer BE, Li R, Barnes M, et al. A global, neutral platform for sharing trial data. *N Engl J Med* 2016;374(25):2411-2413.

Calcagno V, Demoinet E, Gollner K, et al. Flows of research manuscripts among scientific journals reveal hidden submission patterns. *Science* 2012;338:1065-1069.

Camby I, Delpire V, Rouxhet L, et al. Publication practices and standards: recommendations from GSK Vaccines' author survey. *Trials* 2014;15:446.

Chen R, Desai NR, Ross JS, et al. Publication and reporting of clinical trial results: cross sectional analysis across academic medical centers. *BMJ* 2016;352:i637.

Collins S. What are the major sources of capital for medical device companies? November 2015. Accessed http://marketrealist.com/2015/11/major-sources-capital-medical-device-companies/

Committee of Publication Ethics. Code of conduct and best practice guidelines for journal editors. 2011. Accessed http://publicationethics.org/files/Code of conduct for journal editors Mar11.pdf.

Committee of Publication Ethics. Ethical guidelines for peer reviewers. 2013. Accessed http://publicationethics.org/files/Peer%20review%20guidelines 0.pdf.

Council of Science Editors. White paper on promoting integrity in scientific journal publications, 2012 update. Accessed http://www.councilscienceeditors.org/wp-content/uploads/entire whitepaper.pdf

DeTora L, Foster C, Skobe C, et al. Publication planning: promoting an ethics of transparency and integrity in biomedical research. *Int J Clin Pract* 2015;69:915-921.

Drazen JM, Morrissey S, Malina D, et al. The importance—and the complexities—of data sharing. *N Engl J Med* 2016;375(12):1182-1183.

Dwan K, Altman DG, Arnaiz JA, et al. Systematic review of the empirical evidence of study publication bias and outcome reporting bias. *PLoS ONE* 2008;3(8):e3081.

Dwan K, Gamble C, Williamson PR, et al. Systematic review of the empirical evidence of study publication bias and outcome reporting bias—an updated review. *PLoS ONE* 2013;8(7):e66844.

Eurobarometer Science & Technology Report. June 2010. Accessed: http://ec.europa.eu/public_opinion/archives/ebs/ebs_340_en.pdf

Freire D. Peering at open peer review. *The Political Methodist* 2015. Accessed https://thepoliticalmethodologist.com/2015/12/08/peering-at-open-peer-review/

Gardner T & Inger S. How readers discover content in scholarly journals, Summary Edition. Comparing the changing user behavior between 2005 and 2012 and its impact on publisher web site design and function. Accessed http://www.renewtraining.com/how-readers-discover-content-in-scholarly-journals-summary-edition.pdf.

Gattrell WT, Hopewell S, Young K, et al. Professional medical writing support and the quality of randomized controlled trial reporting: a cross-sectional study. *BMJ Open* 2016;6:e010329.

Gauchat G. Politicization of science in the public sphere: a study of public trust in the United States, 1974 to 2010. *Am Sociological Rev* 2012;77(2):167-187.

Goldacre B & Gray J. OpenTrials: towards a collaborative open database of all available information on all clinical trials. *Trials* 2016;17:164.

Groves T & Loder E. Prepublication histories and open peer review at the BMJ. *BMJ* 2014;349:g5394.

Haug CJ. From patient to patient—sharing the data from clinical trials. *N Engl J Med* 2016;374(25):2409-2411.

Hopewell S, Collins GS, Boutron I, et al. Impact of peer review on reports of randomized trials published in open peer review journals: retrospective before and after study. *BMJ* 2014;349:g4145.

Huth EJ & Case K. The URM: twenty-five years old. *Science Editor* 2004; 27:17-21.

Institute of Medicine (IOM). *Sharing clinical trial data: Maximizing benefits minimizing risk.* Washington, DC: The National Academies Press, 2015.

Jacobs a. Adherence to the CONSORT guideline in papers written by professional medical writers. *Write Stuff* 2010;19:196-200.

Jinha A. Article 50 million: an estimate of the number of scholarly articles in existence. *Learned Publishing* 2010;23:258-263.

Jumbe NL, Murray JC, Kern S. Data sharing and inductive learning— toward healthy birth, growth, and development. *N Engl J Med* 2016;374(25):2415-2417.

Joint position on the publication of clinical trial results in the scientific literature. June 2010. Accessed http://www.efpia.eu/uploads/Modules/ Documents/20100610_joint_position_publication_10jun2010.pdf

Knight SR. Social media and online attention as an early measure of the impact of research in solid organ transplantation. *Transplantation* 2014;98(5):490-496.

Krumholz HM & Waldstreicher J. The Yale Open Data Access (YODA) Project—a mechanism for data sharing. *N Engl J Med* 2016;375(5): 403-405.

Laine C & Mulrow CD. Exorcising ghosts and unwelcome guests. *Ann Intern Med* 2005;143(8):611-612.

Lee CJ, Sugimoto CR, Zhang G, et al. Bias in peer review. *J Am Soc Info Sci Tech* 2013;64(1):2-17.

Lee KP, Boyd EA, Holroyd-Leduc JM, et al. Predictors of publication: characteristics of submitted manuscripts associated with acceptance at major biomedical journals. *Med J Aust* 2006;184:621-626.

Lineberry N, Berlin JA, Mansi B, et al. Recommendations to improve adverse event reporting in clinical trial publications: a joint pharmaceutical industry/journal editor perspective. *BMJ* 2016;355:i5078.

Lo B & DeMets DL. Incentives for clinical trialists to share data. *N Engl J Med* 2016;375(12):1112-1115.

Longo DL & Drazen JM. Data sharing. *N Engl J Med* 2016;374(3):276-277.

Marchington JM & Burd GP. Author attitudes to professional medical writing support. *Curr Med Res Opin* 2014;30:2103-2108.

Marušić A, Hren D, Mansi B, et al. Five-step authorship framework to improve transparency in disclosing contributors to industry-sponsored clinical trial publications. *BMC Medicine* 2014;12:197.

Massey PR, Wang R, Prasad V, et al. Assessing the eventual publication of clinical trial abstracts submitted to a large annual oncology meeting. *The Oncologist* 2016;21:261-268.

Merson L, Gaye O, Guerin PJ. Avoiding data dumpsters—toward equitable and useful data sharing. *N Engl J Med* 2016;374(25):2414-2415.

Miller JE, Korn D, Ross JS. Clinical trial registration, reporting, publication and FDAAA compliance: a cross-sectional analysis and ranking of new drugs approved by the FDA in 2012. *BMJ Open* 2015;5:e009758.

Misemer BS, Platts-Mills TF, Jones CW. Citation bias favoring positive clinical trials of thrombolytics for acute ischemic stroke: a cross-sectional analysis. *Trials* 2016;17:473.

Moses H, Matheson DHM, Cairns-Smith S, et al. The anatomy of medical research US and international comparisons. *JAMA* 2015;313(2):174-189.

Navar AM, Pencina MJ, Rymer JA, et al. Use of open access platforms for clinical trial data. *JAMA* 2016;315(12):1283-1284.

Peters I, Kraker P, Lex E, et al. Research data explored: an extended analysis of citations and altmetrics. *Scientometrics* 2016;107:723-744.

PhRMA. A decade of innovation in rare diseases 2005-2015. Accessed http://www.phrma.org/sites/default/files/pdf/PhRMA-Decade-of-Innovation-Rare-Diseases.pdf

PhRMA. 2015 Profile Biopharmaceutical Research Industry. Accessed http://www.phrma.org/sites/default/files/pdf/2015_phrma_profile.pdf

Platt R & Ramsberg J. Challenges for sharing data from embedded research. *N Engl J Med* 2016;374:1897.

Powell K. The waiting game. *Nature* 2016;530:148-151.

Rawal B & Deane BR. Clinical trial transparency: an assessment of the disclosure of results of company-sponsored trials associated with new medicines approved recently in Europe. *Curr Med Res Opin* 2014;30:395-405.

Rockhold F, Nisen P, Freeman A. Data sharing at a crossroads. *N Engl J Med* 2016;375(12):1115-1117.

Ross JS, Tse T, Zarin DA, et al. Publication of NIH funded trials in ClinicalTrials.gov: cross sectional analysis. *BMJ* 2012;344:d7292.

Siler K, Lee K, Bero L. Measuring the effectiveness of scientific gatekeeping. *PNAS* 2015;112(2):360-365.

Taichman DB, Backus J, Baethge C, et al. Sharing clinical trial data—a proposal from the International Committee of Medical Journal Editors. *N Engl J Med* 2016;374:384-386.

The Academic Research Organization Consortium of Continuing Evaluation of Scientific Studies—Cardiovascular (ACCESS CV). Sharing data from cardiovascular clinical trials—a proposal. *N Engl J Med* 2016;375(5):407-409.

The International Consortium of Investigators for Fairness in Trial Data Sharing. Toward fairness in data sharing. *N Engl J Med* 2016; 375(5):405-407.

Thelwall M, Haustein S, Lariviere V, et al. Do altmetrics work? Twitter and ten other social web services. *PLoS ONE* 2013;8(5):e64841.

Van Noorden R. The true cost of science publishing. *Nature* 2013;495:426-429.

Van Rooyen S, Godlee F, Evans S, et al. Effect of open peer review on quality of reviews and on reviewers' recommendations: a randomized trial. *BMJ* 1999;318:23-27.

Van Rooyen S, Delamothe T, Evans SJW. Effect of peer review of telling reviewers that their signed reviews might be posted on the web: randomized controlled trial. *BMJ* 2010;341:c5729.

Walsh E, Rooney M, Appleby L, et al. Open peer review: a randomized controlled trial. *Brit J Psych* 2000;176:47-51.

Ware M & Mabe M. The stm report. An overview of scientific and scholarly journal publishing. September 2009. Accessed http://www.stm-assoc.org/2009 10 13 MWC STM Report.pdf.

Ware M & Mabe M. The stm report. An overview of scientific and scholarly journal publishing. March 2015. Accessed http://www.stm-assoc.org/2015_02_20_STM_Report_2015.pdf.

Warren E. Strengthening research through data sharing. *N Engl J Med* 2016;375(5):401-403.

World Medical Association. Declaration of Helsinki Ethical Principles for Medical Research Involving Human Subjects. October 2008. Accessed http://www.wma.net/en/30publications/10policies/b3/17c.pdf.

Recommended Guidelines

CHEERS: Consolidated Health Economic Evaluation Reporting Standards: http://www.valueinhealthjournal.com/article/S1098-3015(13)00022-3/pdf

CONSORT: Consolidated Standards of Reporting Trials: www.consort-statement.org

COPE: Committee on Publication Ethics: www.publicationethics.org

EQUATOR: Enhancing the Quality & Transparency of Health Research: www.equator-network.org

GPP3: Battisti WP, Wager E, Baltzer L, et al. Good publication practice for communicating company-sponsored medical research: GPP3. *Ann Intern Med* 2015; 163:461-64.: http://annals.org/aim/article/2424869/good-publication-practice-communicating-company-sponsored-medical-research-gpp3

ICMJE: International Committee of Medical Journal Editors (Recommendations): www.icmje.org

Joint Position on the Publication of Clinical Trial Results in the Scientific Literature (IFPMA, EFPIA, JPMA, PhRMA): http://www.efpia.eu/uploads/Modules/Documents/20100610_joint_position_publication_10jun2010.pdf

MPIP: Medical Publishing Insights & Practices. Several published guidelines are available on publication topics such as transparency of clinical research, authorship, and adverse event reporting: www.mpip-initiative.org

PRISMA: Preferred Reporting Items for Systematic Reviews and Meta-Analyses: www.prisma-statement.org

STROBE: Strengthening the Reporting of Observational Studies in Epidemiology: www.strobe-statement.org